CONE PENETRATION TESTING 2018

PROCEEDINGS OF THE 4TH INTERNATIONAL SYMPOSIUM ON CONE PENETRATION TESTING (CPT'18), DELFT, THE NETHERLANDS, 21–22 JUNE 2018

Cone Penetration Testing 2018

Editors

Michael A. Hicks
Section of Geo-Engineering, Department of Geoscience and Engineering,
Faculty of Civil Engineering and Geosciences, Delft University of Technology,
Delft, The Netherlands

Federico Pisanò
Section of Geo-Engineering, Department of Geoscience and Engineering,
Faculty of Civil Engineering and Geosciences, Delft University of Technology,
Delft, The Netherlands

Joek Peuchen
Fugro, The Netherlands

CRC Press is an imprint of the
Taylor & Francis Group, an **informa** business

A BALKEMA BOOK

Cover image courtesy of Fugro © 2018

CRC Press/Balkema is an imprint of the Taylor & Francis Group, an informa business

© 2018 Delft University of Technology, the Netherlands. All rights reserved. Published by Taylor & Francis Group plc.

Typeset by V Publishing Solutions Pvt Ltd., Chennai, India

All rights reserved. No part of this publication or the information contained herein may be reproduced, stored in a retrieval system, or transmitted in any form or by any means, electronic, mechanical, by photocopying, recording or otherwise, without written prior permission from Delft University of Technology, The Netherlands.

Although all care is taken to ensure integrity and the quality of this publication and the information herein, no responsibility is assumed by the publishers nor the editors and authors for any damage to the property or persons as a result of operation or use of this publication and/or the information contained herein.

The full papers of Cone Penetration Testing 2018 are available online via open access: https://www.taylorfrancis.com/.

Published by: CRC Press/Balkema
Schipholweg 107C, 2316 XC Leiden, The Netherlands
e-mail: Pub.NL@taylorandfrancis.com
www.crcpress.com – www.taylorandfrancis.com

ISBN: 978-1-138-58449-5 (Hardback + USB)
ISBN: 978-0-429-50598-0 (eBook)

Table of contents

Preface	xi
Committees	xiii
Sponsors	xv

Keynote papers

Penetrometer equipment and testing techniques for offshore design of foundations, anchors and pipelines *M.F. Randolph, S.A. Stanier, C.D. O'Loughlin, S.H. Chow, B. Bienen, J.P. Doherty, H. Mohr, R. Ragni, M.A. Schneider, D.J. White & J.A. Schneider*	3
Inverse filtering procedure to correct cone penetration data for thin-layer and transition effects *R.W. Boulanger & J.T. DeJong*	25
Use of CPT for the design of shallow and deep foundations on sand *K.G. Gavin*	45

CPT'18 papers

Evaluating undrained rigidity index of clays from piezocone data *S.S. Agaiby & P.W. Mayne*	65
On the use of CPT for the geotechnical characterization of a normally consolidated alluvial clay in Hull, UK *L. Allievi, T. Curran, R. Deakin & C. Tyldsley*	66
New vibratory cone penetration device for in-situ measurement of cyclic softening *D. Al-Sammarraie, S. Kreiter, F.T. Stähler, M. Goodarzi & T. Mörz*	67
Estimation of geotechnical parameters by CPTu and DMT data: A case study in Emilia Romagna (Italy) *S. Amoroso, P. Monaco, L. Minarelli, M. Stefani & D. Marchetti*	68
Cone penetration testing on liquefiable layers identification and liquefaction potential evaluation *E. Anamali, L. Dhimitri, D. Ward & J.J.M. Powell*	69
Dutch field tests validating the bearing capacity of Fundex piles *S. Van Baars, S. Rica, G.A. De Nijs, G.J.J. De Nijs & H.J. Riemens*	70
Estimation of spatial variability properties of mine waste dump using CPTu results—case study *I. Bagińska, M. Kawa & W. Janecki*	71
Strength parameters of deltaic soils determined with CPTU, DMT and FVT *L. Bałachowski, K. Międlarz & J. Konkol*	72
Estimation of the static vertical subgrade reaction modulus k_s from CPT *N. Barounis & J. Philpot*	73
Estimation of in-situ water content and void ratio using CPT for saturated sands *N. Barounis & J. Philpot*	74

NMO-SCTT: A unique SCPT tomographic imaging algorithm E. Baziw & G. Verbeek	75
Effect of piezocone penetration rate on the classification of Norwegian silt A. Bihs, S. Nordal, M. Long, P. Paniagua & A. Gylland	76
Quantifying the effect of wave action on seabed surface sediment strength using a portable free fall penetrometer C. Bilici, N. Stark, A. Albatal, H. Wadman & J.E. McNinch	77
Interpretation of seismic piezocone penetration test and advanced laboratory testing for a deep marine clay M.D. Boone, P.K. Robertson, M.R. Lewis, D.E. Gerken & W.P. Duffy	78
CPT-based liquefaction assessment of CentrePort Wellington after the 2016 Kaikoura earthquake J.D. Bray, M. Cubrinovski, C. de la Torre, E. Stocks & T. Krall	79
Use of CPTu for design, monitoring and quality assurance of DC/DR ground improvement projects A. Brik & P.K. Robertson	80
Cost effective ground improvement solution for large scale infrastructure project A. Brik & D. Tonks	81
Evaluation of existing CPTu-based correlations for the deformation properties of Finnish soft clays B. Di Buò, J. Selänpää, T. Länsivaara & M. D'Ignazio	82
Ultimate capacity of the drilled shaft from CPTu test and static load test M.A. Camacho, J. Mejia, C.B. Camacho, W. Heredia & L.M. Salinas	83
Interpreting properties of glacial till from CPT and its accuracy in determining soil behaviour type when applying it to pile driveability assessments A. Cardoso, S. Raymackers, J. Davidson & S. Meissl	84
Variable rate of penetration and dissipation test results in a natural silty soil R. Carroll & P. Paniagua	85
Rapid penetration of piezocones in sand S.H. Chow, B. Bienen & M.F. Randolph	86
Applying Bayesian updating to CPT data analysis S. Collico, N. Perez, M. Devincenzi & M. Arroyo	87
Geotechnical characterization of a very soft clay deposit in a stretch of road works R.Q. Coutinho, H.T. Barbosa & A.D. Gusmão	88
Analysis of drainage conditions for intermediate soils from the piezocone tests R.Q. Coutinho & F.C. Mellia	89
Behaviour of granitic residual soils assessed by SCPTu and other in-situ tests N. Cruz, J. Cruz, C. Rodrigues, M. Cruz & S. Amoroso	90
Rate effect of piezocone testing in two soft clays F.A.B. Danziger, G.M.F. Jannuzzi, A.V.S. Pinheiro, M.E.S. Andrade & T. Lunne	91
A modified CPT based installation torque prediction for large screw piles in sand C. Davidson, T. Al-Baghdadi, M. Brown, A. Brennan, J. Knappett, C. Augarde, W. Coombs, L. Wang, D. Richards, A. Blake & J. Ball	92
Effects of clay fraction and roughness on tension capacity of displacement piles L.V. Doan & B.M. Lehane	93
Shaft resistance of non-displacement piles in normally consolidated clay L.V. Doan & B.M. Lehane	94

Effects of partial drainage on the assessment of the soil behaviour type using the CPT　95
L.V. Doan & B.M. Lehane

Analysis of CPTU data for the geotechnical characterization of intermediate sediments　96
M.F. García Martínez, L. Tonni, G. Gottardi & I. Rocchi

Detection of soil variability using CPTs　97
T. de Gast, P.J. Vardon & M.A. Hicks

MPM simulation of CPT and model calibration by inverse analysis　98
P. Ghasemi, M. Calvello, M. Martinelli, V. Galavi & S. Cuomo

Challenges in marine seismic cone penetration testing　99
P. Gibbs, R.B. Pedersen, L. Krogh, N. Christopher, B. Sampurno & S.W. Nielsen

Numerical simulation of cone penetration test in a small-volume calibration chamber:
The effect of boundary conditions　100
M. Goodarzi, F.T. Stähler, S. Kreiter, M. Rouainia, M.O. Kluger & T. Mörz

Transition- and thin layer corrections for CPT based liquefaction analysis　101
J. de Greef & H.J. Lengkeek

Soil classification of NGTS sand site (Øysand, Norway) based on CPTU, DMT and
laboratory results　102
A.S. Gundersen, S. Quinteros, J.S. L'Heureux & T. Lunne

Numerical study of anisotropic permeability effects on undrained CPTu penetration　103
L. Hauser, H.F. Schweiger, L. Monforte & M. Arroyo

Evaluating undrained shear strength for peat in Hokkaido from CPT　104
H. Hayashi & T. Yamanashi

Interpreting improved geotechnical properties from RCPTUs in KCl-treated quick clays　105
T.E. Helle, M. Long & S. Nordal

Some experiences using the piezocone in Mexico　106
E. Ibarra-Razo, R. Flores-Eslava, I. Rivera-Cruz & J.L. Rangel-Núñez

Evaluation of complex and/or short CPTu dissipation tests　107
E. Imre, T. Schanz, L. Bates & S. Fityus

New Russian standard CPT application for soil foundation control on permafrost　108
O.N. Isaev, R.F. Sharafutdinov, N.G. Volkov, M.A. Minkin, G.Y. Dmitriev & I.B. Ryzhkov

Thermophysical finite element analysis of thawing of frozen soil by means of HT-CPT
cone penetrometer　109
O.N. Isaev, R.F. Sharafutdinov & D.S. Zakatov

Large deformation modelling of CPT probing in soft soil—pore water pressure analysis　110
J. Konkol & L. Bałachowski

The use of neural networks to develop CPT correlations for soils in northern Croatia　111
M.S. Kovacevic, K.G. Gavin, C. Reale & L. Libric

CPT in thinly inter-layered soils　112
D.A. de Lange, J. Terwindt & T.I. van der Linden

CPT based unit weight estimation extended to soft organic soils and peat　113
H.J. Lengkeek, J. de Greef & S. Joosten

Impact of sample quality on CPTU correlations in clay—example from the Rakkestad clay　114
*J.S. L'Heureux, A.S. Gundersen, M. D'Ignazio, T. Smaavik, A. Kleven, M. Rømoen,
K. Karlsrud, P. Paniagua & S. Hermann*

Use of the free fall cone penetrometer (FF-CPTU) in offshore landslide hazard assessment　115
J.S. L'Heureux, M. Vanneste, A. Kopf & M. Long

Fibre optic cone penetrometer P. Looijen, N. Parasie, D. Karabacak & J. Peuchen	116
Some considerations related to the interpretation of cone penetration tests in sulphide clays in eastern Sweden A.B. Lundberg & E.A. Alderlieste	117
Effect of cone penetrometer type on CPTU results at a soft clay test site in Norway T. Lunne, S. Strandvik, K. Kåsin, J.S. L'Heureux, E. Haugen, E. Uruci, A. Veldhuijzen, M. Carlson & M. Kassner	118
Evaluation of CPTU N_{kt} cone factor for undrained strength of clays P.W. Mayne & J. Peuchen	119
Applying breakage mechanics theory to estimate bearing capacity from CPT in polar snow A.B. McCallum	120
Empirical correlations to improve the use of mechanical CPT in the liquefaction potential evaluation and soil profile reconstruction C. Meisina, S. Stacul & D.C. Lo Presti	121
Rigidity index (I_R) of soils of various origin from CPTU and SDMT tests Z. Młynarek, J. Wierzbicki & K. Stefaniak	122
A state parameter-based cavity expansion analysis for interpretation of CPT data in sands P.Q. Mo & H.S. Yu	123
Permeability estimates from CPTu: A numerical study L. Monforte, M. Arroyo, A. Gens & C. Parolini	124
Pore pressure measurements using a portable free fall penetrometer M.B. Mumtaz, N. Stark & S. Brizzolara	125
Influence of soil characteristics on cone and ball strength factors: Case studies T.D. Nguyen & S.G. Chung	126
A method for predicting the undrained shear strength from piezocone dissipation test E. Odebrecht, F.M.B. Mantaras & F. Schnaid	127
Realistic numerical simulations of cone penetration with advanced soil models Z.Y. Orazalin & A.J. Whittle	128
Calibrating NTH method for ϕ' in clayey soils using centrifuge CPTu Z. Ouyang & P.W. Mayne	129
CPT interpretation and correlations to SPT for near-shore marine Mediterranean soils S. Papamichael & C. Vrettos	130
Characterization of a dense deep offshore sand with CPT and shear wave velocity profiling J.G. Parra, A. Veracoechea & N. Vieira	131
Calibration of cone penetrometers in accredited laboratory J. Peuchen, D. Kaltsas & G. Sinjorgo	132
Defining geotechnical parameters for surface-laid subsea pipe-soil interaction J. Peuchen & Z. Westgate	133
Shallow depth characterisation and stress history assessment of an over-consolidated sand in Cuxhaven, Germany V.S. Quinteros, T. Lunne, L. Krogh, R. Bøgelund-Pedersen & J. Brink Clausen	134
Assessment of pile bearing capacity and load-settlement behavior, based on Cone Loading Test (CLT) results Ph. Reiffsteck, H. van de Graaf & C. Jacquard	135
CPT based settlement prediction of shallow footings on granular soils J. Rindertsma, W.J. Karreman, S.P.J. Engels & K.G. Gavin	136

Analysis of acceleration and excess pore pressure data of laboratory impact penetrometer tests in remolded overconsolidated cohesive soils   137
R. Roskoden, A. Kopf, T. Mörz & S. Kreiter

CPT-based parameters of pile lengths in Russia   138
I.B. Ryzhkov & O.N. Isaev

Comparison of settlements obtained from zone load tests and those calculated from CPT and PMT results   139
A. Sbitnev, B. Quandalle & J.D. Redgers

Direct use of CPT data for numerical analysis of VHM loading of shallow foundations   140
J.A. Schneider, J.P. Doherty, M.F. Randolph & K. Krabbenhøft

Evaluation of existing CPTu-based correlations for the undrained shear strength of soft Finnish clays   141
J. Selänpää, B. Di Buò, M. Haikola, T. Länsivaara & M. D'Ignazio

Applications of RCPTU and SCPTU with other geophysical test methods in geotechnical practice   142
Z. Skutnik, M. Bajda & M. Lech

CPT in a tropical collapsible soil   143
C.S.M. Soares, F.A.B. Danziger, G.M.F. Jannuzzi, I.S.M. Martins & M.E.S. Andrade

Liquefaction resistance by static and vibratory cone penetration tests   144
F.T. Stähler, S. Kreiter, M. Goodarzi, D. Al-Sammarraie & T. Mörz

In situ characterisation of gas hydrate-bearing clayey sediments in the Gulf of Guinea   145
F. Taleb, S. Garziglia & N. Sultan

Gas effect on CPTu and dissipation test carried out on natural soft-soil of Barcelona Port   146
D. Tarragó & A. Gens

Comparison of cavity expansion and material point method for simulation of cone penetration in sand   147
F.S. Tehrani & V. Galavi

Soil behavior and pile design: Lesson learned from some prediction events—part 1: Aged and residual soils   148
G. Togliani

Soil behavior and pile design: Lesson learned from recent prediction events—part 2: Unusual NC soils   149
G. Togliani

A probabilistic approach to CPTU interpretation for regional-scale geotechnical modelling   150
L. Tonni, M.F. García Martínez, I. Rocchi, S. Zheng, Z.J. Cao, L. Martelli & L. Calabrese

CPTu-based soil behaviour type of low plasticity silts   151
L.A. Torres-Cruz & N. Vermeulen

Interpretation of soil stratigraphy and geotechnical parameters from CPTu at Bhola, Bangladesh   152
Z.A. Urmi & M.A. Ansary

Thermal Cone Penetration Test (T-CPT)   153
P.J. Vardon, D. Baltoukas & J. Peuchen

Development of numerical method for pile design to EC7 using CPT results   154
J.O. Vasconcelos, J. O'Donovan, P. Doherty & S. Donohue

Prehistoric landscape mapping along the Scheldt by camera- and conductivity CPT-E   155
J. Verhegge, Ph. Crombé & M. van den Wijngaert

Comparative analysis of liquefaction susceptibility assessment by CPTu and SPT tests   156
A. Viana da Fonseca, C. Ferreira, A.S. Saldanha, C. Ramos & C. Rodrigues

Application of CPT testing in permafrost 157
N.G. Volkov, I.S. Sokolov & R.A. Jewell

Comparison of mini CPT cone (2 cm²) vs. normal CPT cone (10 cm² or 15 cm²) data,
2 case studies 158
G.T. de Vries, C. Laban & E. Bliekendaal

The development of "Push-heat", a combined CPT-testing/thermal conductivity
measurement system 159
G.T. de Vries & R. Usbeck

Free fall penetrometer tests in sand: Determining the equivalent static resistance 160
D.J. White, C.D. O'Loughlin, N. Stark & S.H. Chow

The variability of CPTU results on the AMU-Morasko soft clay test site 161
J. Wierzbicki, R. Radaszewski & M. Waliński

Shear strengths determined for soil stability analysis using the digital Icone Vane 162
M. Woollard, O. Storteboom, A.S. Damasco Penna & É.S.V. Makyama

Metal objects detected and standard parameters measured in a single CPT using the Icone
with Magneto click-on module 163
M. Woollard, O. Storteboom, L. Gosnell & P. Baptie

Simulation of liquefaction and consequences of interbedded soil deposits using CPT data 164
F. Yi

Correlations among SCPTU parameters of Jiangsu normally consolidated silty clays 165
H. Zou, S. Liu, G. Cai & A.J. Puppala

Author index 167

Preface

The Technical Committee TC102 of the International Society for Soil Mechanics and Geotechnical Engineering (ISSMGE) and the Organising Committee of CPT'18 extend a warm welcome to all participants of the 4th International Symposium on Cone Penetration Testing (CPT'18). The symposium is being held at Delft University of Technology, in Delft, The Netherlands, 21–22 June 2018, and is the fourth in a series of successful symposia endorsed by TC102 of the ISSMGE, following on from: Linköping, Sweden, 1995; Huntington Beach, California, 2010; and Las Vegas, Nevada, 2014.

The Netherlands has a long history of pioneering research in geotechnical engineering, soil mechanics and testing of soils, motivated by the necessity for engineering constructions on reclaimed land and problematic deltaic soils. This included the first application of (electric, strain gauge) cone penetration testing by Fugro in 1964. Fugro then hosted the first symposium on cone penetration testing in 1972 in The Netherlands, with the proceedings of this "Sondeer Symposium 1972" comprising 6 papers in the Dutch language. One of the highlights of this 2018 symposium is a paper on a fibre optic cone penetrometer—a probable first. It is therefore with great pleasure that Delft University of Technology is hosting CPT'18.

The aim of the symposium is to provide a forum for exchange of ideas and discussion on topics related to the application of cone penetration testing in geotechnical engineering. In particular, the symposium will focus on the solution of geotechnical challenges using the cone penetration test (CPT), CPT add-on measurements and companion in-situ penetration tools (such as full flow and free fall penetrometers), with an emphasis on practical experience and application of research findings. As CPTs play a major role in geotechnical engineering, CPT'18 will bring together the world's experts who are working to improve the quality of cone penetration testing and reduce the difficulties involved. The symposium will be attended by academics, researchers, consultants, practitioners, hardware/software suppliers, certifiers and students. It will be a unique opportunity for meeting people and sharing high-level knowledge.

The peer-reviewed papers contained in the proceedings have been authored by academics, researchers and practitioners from many countries worldwide and cover numerous important aspects related to cone penetration testing in geotechnical engineering, ranging from the development of innovative theoretical and numerical methods of interpretation, to real field applications. The main topics are: interpretation of CPT results, application of the CPT, and equipment development.

A total of 189 abstracts were submitted and the authors of those abstracts that fell within the scope of the symposium were invited to submit full papers for peer review. A total of 122 papers were received and, of these, 104 papers were accepted for inclusion in the symposium proceedings (including 3 keynote papers). The editors would like to thank the Scientific Committee for their assistance in the review process. They would also like to thank the keynote lecturers, authors, participants and sponsors. The editors are grateful for the support of the chair and members of TC102.

On behalf of the Organising Committee, we welcome you to The Netherlands and hope that you find the symposium both enjoyable and inspiring.

Michael Hicks
Federico Pisanò
Joek Peuchen
April 2018

Cone Penetration Testing 2018 – Hicks, Pisanò & Peuchen (Eds)
© 2018 Delft University of Technology, The Netherlands, ISBN 978-1-138-58449-5

Committees

ORGANISING COMMITTEE

Michael Hicks, *Delft University of Technology, The Netherlands (Chair)*
Federico Pisanò, *Delft University of Technology, The Netherlands (Co-Chair)*
Joek Peuchen, *Fugro, The Netherlands (Co-Chair)*
Matthijs Baurichter, *Delft University of Technology, The Netherlands*

SCIENTIFIC COMMITTEE

Shiaohuey Chow, *The University of Western Australia, Australia*
Jason DeJong, *University of California, Davis, USA*
David Frost, *Georgia Institute of Technology, USA*
Jürgen Grabe, *Hamburg University of Technology, Germany*
Michael Jefferies, *Golder Associates, UK*
Cristina Jommi, *Delft University of Technology, The Netherlands and Politecnico di Milano, Italy*
Mandy Korff, *Deltares and Delft University of Technology, The Netherlands*
Lone Krogh, *Ørsted Wind Power, Denmark*
Michael Long, *University College Dublin, Ireland*
Dirk Luger, *Deltares, The Netherlands*
Tom Lunne, *Norwegian Geotechnical Institute, Norway*
Diego Marchetti, *Studio Prof. Marchetti, Italy*
Paul Mayne, *Georgia Institute of Technology, USA*
Dominique Ngan-Tillard, *Delft University of Technology, The Netherlands*
Eric Parker, *RINA Consulting, Italy*
John Powell, *Geolabs Ltd., UK*
Nick Ramsey, *Fugro, Australia*
Peter Robertson, *Gregg Drilling and Testing, USA*
Rodrigo Salgado, *Purdue University, USA*
Nina Stark, *Virginia Tech, USA*
Nabil Sultan, *Ifremer, France*
Frits van Tol, *Deltares, The Netherlands*
Jędrzej Wierzbicki, *Adam Mickiewicz University in Poznań, Poland*
Cor Zwanenburg, *Deltares, The Netherlands*

Sponsors

Symposium Organiser

Gold Sponsors

Silver Sponsors

Bronze Sponsors

Keynote papers

Penetrometer equipment and testing techniques for offshore design of foundations, anchors and pipelines

M.F. Randolph, S.A. Stanier, C.D. O'Loughlin, S.H. Chow, B. Bienen, J.P. Doherty, H. Mohr, R. Ragni & M.A. Schneider
University of Western Australia, Perth, Australia

D.J. White
University of Southampton, Southampton, UK

J.A. Schneider
USACE, St. Paul, MN, USA

ABSTRACT: This paper attempts to categorise geotechnical field site characterisation tools in a hierarchical manner, as appropriate for the progression from initial surveys to detailed geotechnical design of specific infrastructure. In general, the hierarchy reflects more the sophistication, and hence cost, of the field tools, although small-scale tools developed to explore box core samples are something of an exception, with the potential for high quality data at low cost. These ideas are explored in the context of modern developments in equipment and methods of deployment, and in the manner in which the data may be used efficiently in design.

1 INTRODUCTION

The current economic climate for offshore energy is requiring improved efficiency in the design of subsea infrastructure, a key starting point for which is optimising site investigation (SI) data and their application in design. Geotechnical design for deep water developments is dominated by subsea infrastructure such as subsea foundation units and in-field pipelines founded within the uppermost sediment layers. Even anchoring systems, particularly for mobile drilling systems, will generally lie within the top 10 to 20 m of the seabed. Hence the vast majority of deep water infrastructure is founded within sediments that are well within reach of relatively modest-sized robotic seabed tools or free-fall penetrometers. There is, however, a marked difference between testing requirements for pipeline design parameters, where the focus is generally the upper 0.5 m of the seabed, and those for subsea foundations and anchors.

For coastal and other shallow water developments, the challenge of acquiring in situ test data is reduced somewhat, although for applications like cable laying it may be sufficient just to identify sediment type. Economic pressures, particularly for wind and wave energy applications, will require optimisation of the spatial frequency of acquiring data in situ, and how the data are then applied in design to best advantage.

This paper discusses advances in site investigation approaches, with initial focus on novel approaches for near-surface characterisation and cost-effective free-fall penetrometry, which have particular relevance for deep water projects or for identifying surficial sediment types generally. The advances are considered in light of the need to balance cost and quality of data at different phases of a project. The paper also discusses the growth in 'direct' use of CPT data in design methods, such as for jack-up rig foundations. Potential differences between 'indirect' use of CPT data to derive simplified strength profiles, and more direct incorporation of what may appear relatively minor fluctuations in the cone resistance through the stratigraphy, are examined.

2 PHASES OF GEOTECHNICAL SITE CHARACTERISATION

2.1 *Exploratory phases*

For a new development, early investigation phases are focused on obtaining geophysical data in order to establish a broad geological model for the region. Historically, very limited (if any) geotechnical data on the upper seabed sediments are obtained until later. This approach is influenced to a large degree by cost, since geophysical data may be gathered using relatively small vessels, with limited deck space for accommodating geotechnical investigation equipment.

Nowadays, however, lightweight (compact) tools, such as box core samplers, free-fall penetrometers and lightweight seabed frames equipped with small diameter coiled rod cone penetrometers are being incorporated increasingly in geophysical investigations. These provide early indications of the near-surface sediment properties, which are difficult to assess from standard seismic reflection tools (Peuchen & Westgate 2018).

In spite of inevitable disturbance of the soil recovered in a box core, reasonable quantitative estimates of intact and remoulded shear strength may be obtained from miniature penetrometer (T-bar or ball) tests conducted in the upper 0.3 m or so, depending on sample recovery (Low et al. 2008). As discussed in more detail later, additional parameters for pipeline design may be determined from torsional tests on novel types of penetrometer (White et al. 2017).

Free-fall penetrometers also offer cost-effective investigation of the near surface seabed. Even simple devices that just measure the deceleration (e.g. Stark et al. 2009, 2012) will allow approximate estimation of the strength of the upper sediments, providing useful guidance for pipeline route assessment. Modern lightweight free-fall piezocones offer much more quantitative assessment of the strength profiles through the upper material (Stegmann et al. 2006), provided the interpretation approach is appropriate (Chow et al. 2017).

2.2 Main geotechnical SI phases

In the main phases of geotechnical site investigation, larger vessels capable of handling larger and more sophisticated drilling and field testing equipment are used. Even here, though, the modern trend is for more compact, seabed-based robotic drilling and testing equipment, rather than conventional drill ships. In deep water, seabed-based systems are preferable in terms of cost (with compact equipment allowing smaller vessels) and datum stability since the systems are decoupled from wave-induced vessel motions.

Advances in robotic control have led to a number of commercial seabed-based robotic drilling, sampling and testing systems. The pioneer amongst these in terms of increased depth of water and soil depth investigated was the portable remotely operated drill (PROD), developed by Benthic Geotech (Figure 1). The equipment fits within standard shipping containers for transport.

Figure 1. Benthic Geotech's portable remotely operated drill (PROD).

Figure 2. Fugro seafloor drill.

PROD and other systems, such as the Fugro Seafloor Drill (Figure 2), use an umbilical cable to power the seabed frame, although smaller systems can use remotely operated vehicles (ROVs) as a power source (Randolph 2016).

Seabed frames with advanced actuation offer the potential for more sophisticated penetration testing, for example by varying the rate of penetration in order to explore effects of strain rate and partial consolidation. This can yield measurements that are more representative of conditions around the target infrastructure—which may involve significant changes in soil strength due to installation and in-service loading.

Full-flow penetrometers, such as the T-bar and piezoball with projected areas typically a factor of 10 greater than that of the shaft immediately behind the probe, provide potential for cyclic motion in order to remould the surrounding soil. They are often considered as applicable only for shallow uniform fine-grained sediments. However, experience in layered carbonate material has shown that they provide consistent data to CPTs in layered sediments, generally with T-bar and ball resistances showing slightly lower magnitudes of spikes in resistance (see Figure 3).

The addition of pore pressure sensors on so-called 'piezoball' penetrometers, either as discrete button filters or continuous annular filters, enhances their potential as a stratification tool, and also allows field assessment of consolidation properties by means of dissipation tests (Mahmoodzadeh et al. 2015).

As illustration of this potential, Figure 4 shows the range of penetration resistance mobilised in a carbonate silt, relative to a standard rate penetration test. The data are from the RIGSS JIP (White et al. 2017) and span a factor of 20 between fully remoulded conditions and those after full consolidation.

The data from Figure 4 were obtained under laboratory conditions, in a centrifuge model test, although similar data may be acquired from in situ tests. In the carbonate silt, penetration under undrained conditions even at the very high rates associated with free-fall penetrometers, gives lower resistance than for drained conditions at a very slow penetration rate. In stronger soils, such as medium dense silica sand, undrained conditions will most likely provide the highest penetration resistance due to the dilative nature of such soils. The effect of sand density on the relative magnitudes of drained and undrained penetration resistances is well illustrated by the penetrometer data presented by Chow et al. (2018).

The very high resistances in sand mobilised under undrained conditions have been observed in field free-fall penetrometer tests, leading to rapid arrest of the penetrometer (Stark et al. 2012). Although this limits the depth that can be

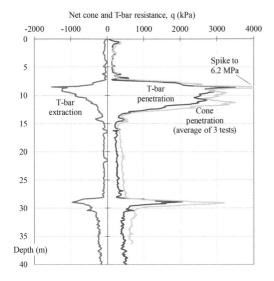

Figure 3. Comparison of in situ cone and T-bar penetration tests from Australian North-West Shelf.

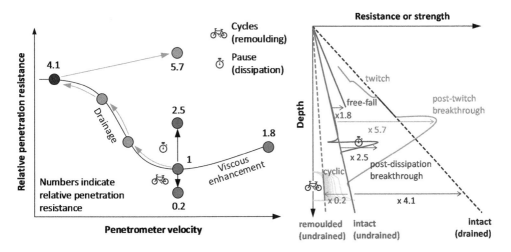

Figure 4. Penetration resistances mobilised in a carbonate silt through different penetrometer test procedures.

explored, the information of shallow sandy sediments with high (undrained) penetration resistance has value for design. With some additional inputs, for example knowledge of mineralogy and particle size of the sediments, the deduced penetration resistances may be interpreted in terms of relative density, in the same way as for static CPTs (White et al. 2018).

Prior to cavitation the undrained strength of sand may be estimated using the framework of Bolton (1986) or a state parameter approach (Been et al. 1991). After cavitation, although this is unlikely to occur in deep water, an upper limit to the penetration resistance may be estimated by considering the cavitation pressure as a 'back pressure' acting in conjunction with the geostatic effective stress state.

Improvements in control and testing procedures open the door to much more sophisticated seabed testing. Figure 5 shows schematically the range of shear strength data that may be acquired, relative to strengths relevant for design of different offshore infrastructure. The diagram reflects the variations in penetration resistance arising from different testing procedures, as discussed with respect to Figure 4. The horizontal spread reflects changes in strength due to consolidation (left side) or disturbance and remoulding (right side), while the vertical axis signifies changes in strength due to shear strain rate.

Modern robotic systems also have the control potential to allow seabed testing that targets stiffness measurement. For example, fatigue design of riser systems requires information on the vertical riser-soil stiffness, and how that varies with time and displacement amplitude of cyclic perturbations.

For a steel catenary riser, the relevant stiffness may vary by 1 or 2 orders of magnitude through the touchdown zone as the displacement amplitude of the riser changes. At a given amplitude, the stiffness will also tend to reduce in the short term as the soil responds to cyclic shearing, but to increase with time due to consolidation (Clukey et al. 2017, Yuan et al. 2017).

Future potential for field tests, either on the seabed or in recovered box core samples (Kelleher et al. 2010, Boscardin & DeGroot 2015), is discussed in the following section, drawing on recent experience on Australia's North-West Shelf and in the Caspian Sea.

3 NOVEL SHALLOW PENETROMETERS FOR PIPELINE DESIGN PARAMETERS

3.1 Background—RIGSS JIP

The Remote Intelligent Geotechnical Seabed Surveys Joint Industry Project (RIGSS JIP) was initiated by the University of Western Australia (UWA) and is reaching the end of the 3 year program of activities. A focus of the JIP has been the development of new and improved techniques for in-situ geotechnical investigation of the upper few metres of the seabed.

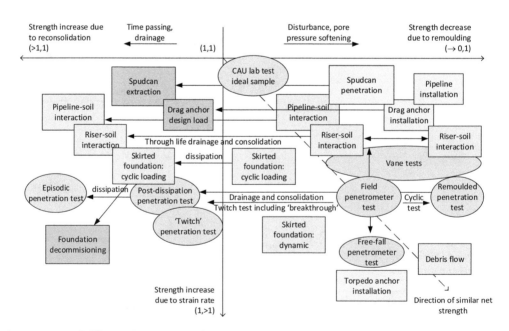

Figure 5. Map of different soil tests and design applications (extended from Randolph et al. 2007).

The RIGSS JIP has pursued three 'visions', which are illustrated in Figure 6:

1. The RIGSS SI: A remote SI platform with robotic control of intelligent tools gathering parameters targeted towards design.
2. Earlier geotechnical definition in projects: leading to reduced uncertainty and risk, and a reduced need to carry multiple options through design.
3. Direct geotechnical design: a design philosophy using in situ test data more directly to determine the response of pipelines, foundations, and other geotechnical systems

Together, these three ideas aim to allow more accurate forecasting of the behaviour of seabed infrastructure. New types of penetrometer can be shaped and manipulated in ways that more closely represent the infrastructure being designed. The resulting data provide the potential to eliminate or simplify the journey from measured resistance to a lab-based strength and back to a predicted resistance or capacity.

A particular focus of the RIGSS JIP has been on shallow surveys that cover extensive areas, such as for pipelines, where low-cost, remotely operated SI tools and smart testing techniques offer potential to improve efficiency and deliver more valuable data. Intelligent tools—such as the shallow penetrometers—give data that can be fed more directly into geotechnical design, reducing the timescale and uncertainty associated with the design outcome.

3.2 Offshore box core testing

Three types of 'shallow' penetrometers have been pursued within the RIGSS JIP. The toroidal and hemispherical ('hemiball') penetrometers were first described by Yan et al. (2011); more recently a ring-shaped variant on the toroid with a flat interface has been trialled. During a test, these devices are penetrated to a depth of less than a diameter (of toroidal bar, or hemiball) and are then subjected to one or more stages of rotation under maintained vertical load. Their shape and the interfacial sliding mode of failure resemble a pipeline, or other surface infrastructure such as mattresses or surface foundations.

Devices at box core scale have been developed and trialled in the UWA laboratory on reconstituted kaolin and carbonate silt samples. The system has also been taken offshore, for field trials on two surveys by the RIGSS JIP partners. The general design of the box core toroid penetrometer and the integrated data acquisition system are shown in Figure 7. The vertical force and torque on the penetrometer are monitored continuously, as are the pore pressures at multiple locations on the surface.

The soil-interface resistance is derived from the measurements, in terms of both total and effective

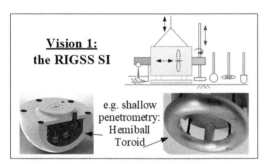

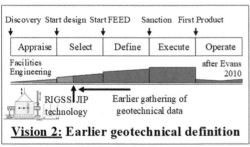

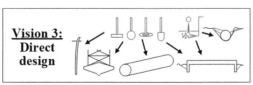

Figure 6. RIGSS JIP motivations (after White et al. 2017).

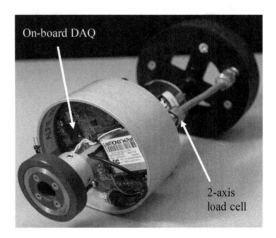

Figure 7. Toroid penetrometer with on-board DAQ and 2-axis load cell (vertical force and torque).

normal stresses. Dissipation stages allow the consolidation coefficient to be determined for fine-grained soils, while the timing of rotation stages may be varied to deduce both drained and undrained interface strengths.

Box-coring operations during one of the recent offshore trials are shown in Figure 8. By utilising multiple box core sleeves, sampling operations may be continued in parallel with shallow penetrometer testing within a previously recovered box core sample. Depending on the drainage properties of the samples, a full set of penetrometer tests in the box core might take 1–3 hours. If this aligns with the cycle time for the vessel transit between locations and box core recovery, then two sleeves can be alternated between sampling and testing.

The box core actuator system is shown during testing within a box core sleeve in Figure 9. Another approach is to take large diameter tube samples from the box cores, and perform tests off the sampling critical path using the arrangement shown in Figure 10.

The initial field trials provided valuable lessons to improve the procedures for deploying the tests and determining the best combination of tests to perform in a given soil type, to optimise the information gathered. The deck of a survey vessel is a challenging environment for performing precision testing, but vessel motions and changes in ambient temperature have had no detrimental effect on the

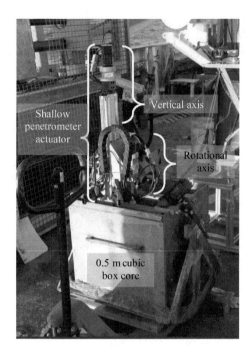

Figure 9. Shallow penetrometer system in 'box core' mode.

Figure 8. Typical box corer during sample recovery to deck.

Figure 10. Shallow penetrometer system in 'tube sample' mode.

measurement quality emerging from the shallow penetrometer tests. To date, the testing equipment has required operation by a researcher familiar with the equipment and software controls. As the technology evolves, simpler systems will be created. We anticipate that shallow penetrometer testing could then be available on a basis similar to the existing Fugro DECKSCOUT and other box core penetrometer systems.

A typical set of results is used to illustrate the performance of the toroid penetrometer (Figure 7), operated in a box core sample of carbonate silt. The initial stage of a toroid test involves vertical penetration of the device to a depth of typically 20% of the diameter (Figure 11). The results are converted to a linear profile of undrained strength, s_u, using the bearing capacity model set out by Stanier & White (2015). In this approach, the bearing factor, N_c, is varied according to the depth and strength gradient, and a (small) adjustment is also made for soil buoyancy. An iterative process is required to reach s_u because N_c varies with s_u itself. The minimal noise in the measured data is equivalent to less than 0.1 kPa of strength (Figure 11).

After reaching the target depth, the vertical load on the penetrometer is maintained constant, while the excess pore pressure at the invert dissipates (Figure 12). The initial response—over the

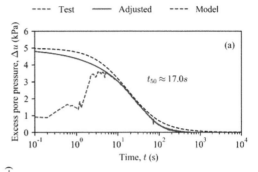

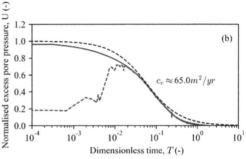

Figure 12. Toroid dissipation in carbonate silt: (a) measured response; and (b) interpreted coefficient of consolidation, c_v.

first five seconds of the dissipation—shows a rise in pore pressure, typical for dilatant soils. During the subsequent 10 minutes, the response follows closely the analytical model set out by Yan et al. (2017), based on numerical analysis. By matching the curves at the 50% dissipation point, a consolidation coefficient of 65 m²/year is determined.

The next stage of the test involves rotation of the toroid (Figure 13). Initially a high rotation rate is used to mobilise the undrained interface strength, including any initial peak but continuing towards the stable or residual value used in pipeline design. The rotation rate is then reduced by a factor of 10. Rotation continues as the resistance rises towards the drained limit, controlled by the interface friction angle.

The measured torque and vertical force are converted to interface friction ratio by dividing the torque by the effective radius (at which the resultant of the circumferential resistance acts) and multiplying the vertical force by a 'wedging factor' (White & Randolph 2007) to give the total normal force on the interface. The resulting interface friction ratio can be fitted by undrained and drained limits (Figure 13), which are key inputs to pipeline design. The rate of transition provides another indication of the consolidation coefficient, which may differ from the initial dissipation stage.

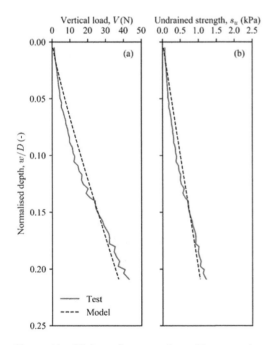

Figure 11. Fitting of measured toroid penetration response in carbonate silt with analytical model: (a) measured vertical load; and (b) interpreted profile of undrained shear strength, s_u.

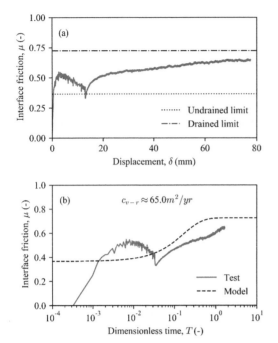
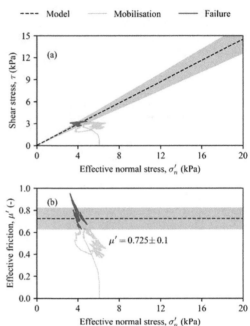

Figure 13. Toroid rotation in carbonate silt: evolution of interface friction, μ with (a) displacement; and (b) dimensionless time.

Figure 14. Toroid effective stress interpretation: (a) effective stress envelope; and (b) effective interface friction, μ'.

The undrained limit, if the soil has been normally consolidated under the penetrometer vertical load, is the 'c/p' or normally consolidated undrained strength ratio for the interface. The drained limit is $\tan(\delta_{res})$ where δ_{res} is the residual interface friction angle. This interpretation is aligned with conventions for pipeline design (White et al. 2017, DNVGL 2017).

The interface response can also be interpreted in effective stress terms using the pore pressure measurements (Figure 14). This provides useful confirmation of the drained friction, particularly if the tests are curtailed before pore pressure equilibrium is reached.

An additional type of test involves episodic rotations with intervening dissipations, which results in gradually increasing undrained resistance with each cycle (Yan et al. 2014, Boscardin & DeGroot 2015). The data may be interpreted within a consolidation framework, and used to support assessment of 'consolidation hardening' by which pipeline axial friction can rise through cycles of movement.

Overall, subject to a suitable box core sample being recoverable, a short series of shallow penetrometers tests (e.g. 2–3 tests performed over several hours), supplemented with conventional miniature T-bar (or ball) penetrometer tests, can provide all of the geotechnical strength and consolidation parameters required for pipeline design. We hope that these technologies will be taken forward by the RIGSS JIP partners over the coming years, allowing wider adoption in project practice.

4 FREE-FALL PENETROMETERS

4.1 Introduction

Free fall penetrometers (FFPs) may be divided into different classes, according to the sophistication of instrumentation (ranging from just accelerometers to fully instrumented with tip and shaft load cells and pore pressure sensors). Recent developments have included two-stage combined FFP systems where free-fall penetration is followed by static penetration. In addition, extended base FFPs have been explored for soft sediments, as discussed later. The discussion below describes different types of FFPs, and then summarises interpretation in fine-grained sediments.

4.2 Probe geometry

A variety off FFPs are shown in Figure 15. They vary in mass and geometry from the 52 mm diameter by 0.215 m long 0.7 kg expendable bottom penetrometer (XBP, Stoll & Akal 1999) to the

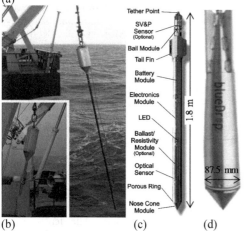

Figure 15. FFPs: (a) CPT Stinger (Young et al. 2011) (b) University of Bremen FFCPT shown in long and short (inset) modes (Stegman et al. 2006) (c) Brooke Ocean 'FFCPT-660' (Mosher et al. 2007) and (d) BlueDrop (Stark et al. 2014).

160 mm diameter (with a standard CPT tip) by 23 to 35 m long 3200 kg CPT-Stinger (Young et al. 2011). Penetration is generally proportional to mass (or momentum) per unit area (Peuchen et al. 2017), with the XBP generally penetrating less than 0.3 m, and the CPT-Stinger capable of reaching 20 m of dynamic penetration followed by another 15 m of static penetration.

A class of intermediate size probes capable of 4 to 6 m of penetration in soft clays, 1 to 2 m penetration in stiff clays and silts, and less than 0.5 m penetration in sands is seeing increased use for shallow investigations in soft materials. These penetrometers are relatively portable and can be used from survey size vessel with a high-speed winch for deployment and retrieval, with diameters of 100 to 200 mm, lengths of 1 to 6 m, and mass of 10 to 500 kg. They include:

- NAVFAC eXpendable Doppler Penetrometer (XDP) (Beard 1985, Orenberg et al. 1996, Thompson et al. 2002)
- Seabed Terminal Impact Naval Gauge (STING) (e.g. Mulhearn 2003)
- Brooke Ocean Free Fall CPT (FFCPT) (Furlong et al. 2006)
- Bremen Free Fall CPT (FFCPT) (Stegmann et al. 2006)
- NIMROD (Stark et al. 2009)
- LIRmeter (Stephan et al. 2011)
- BlueDrop (Stark et al. 2014)
- Graviprobe (Geirnaert et al. 2013)
- Instrumented Free-Fall Sphere (IFFS) (Morton et al. 2016a)
- SEADART1 (Peuchen et al. 2017)

4.3 *Sensors*

Historically FFPs were only equipped for acceleration or velocity measurements. With the need for higher quality measurements of the location of the seabed, soil strength, as well as inferences of soil type, additional sensors such as optical sensors, tip stress, sleeve friction, penetration pore pressure, and resistivity have been added to probes.

The most efficient means of FFP operation is whereby the instrumentation package has sufficient power and storage for a series of tests. In a continuous test mode, where the penetrometer is repeatedly deployed and retrieved, the instrumentation package should have storage and battery capabilities for about 12 hours of testing. After this the package can be removed from the penetrometer for battery recharging and data download, or alternatively the package could be replaced with a spare system, allowing testing to continue without delay.

Modern data acquisition systems allow for this mode of operation. For example, the UWA instrumentation package shown in Figure 16 is sufficiently compact to fit within the smallest FFPs, but allows for sampling rates of up to 500 kHz and at a practical sampling rate of 1.5 kHz for FFP tests, up to 225 hours of data storage and 20 hours of continued use before the batteries require recharging.

All FFPs measure acceleration along an axis aligned with the body of the FFP (although often sensors with different measurement ranges are included to optimise resolution). The UWA instrumentation package shown in Figure 16 is an inertial measurement unit (IMU), which measures acceleration along each axis of a Cartesian coordinate system and provides independent measurements of the rotations about each axis of the same coordinate system. Transforming the measurements from the body reference frame of the FFP (which will be changing during an FFP test if the

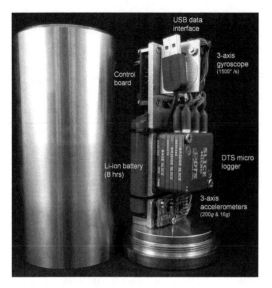

Figure 16. UWA inertial measurement unit (IMU).

FFP tilts by varying amounts) to a fixed reference frame avoids the difficulty in identifying the component of the acceleration signal that is due to FFP rotation from that due to changing acceleration (Blake et al. 2016).

O'Loughlin et al. (2014) showed that the corrections using this approach become appreciable at tilts in excess of 10°. Conical shaped penetrometers appear to tilt less than this threshold (Kopf et al. 2007), but for other penetrometers, particularly the free-fall sphere penetrometer (described later), the tilts are likely to be much higher, such that an IMU approach is required.

4.4 Interpretation using acceleration data

The ultimate objective of a FFP test in fine-grained material is to assess the undrained shear strength s_u. This may be derived from the tip resistance in the conventional way as for a static CPT, but with appropriate allowance for the high strain rates resulting from the high velocity penetration. Ideally, the FFP should measure the tip resistance directly. However, for FFPs that do not include a tip load cell, estimation of the tip resistance involves a number of steps and uncertainties.

The tip resistance may be derived from the penetrometer mass, m, times the measured acceleration, a = v(dv/dz) according to

$$q_{t,FFP} = \frac{W_b - mv(dv/dz) - Q_s - F_D - F_b}{A_{tip}} \quad (1)$$

where v is the FFP velocity, z the penetration, W_b the submerged weight of the FFP in water, Q_s the frictional resistance along the shaft, F_D the drag resistance (due to soil inertia), F_b the buoyancy force equal to the effective weight of the displaced soil and A_{tip} the cross-sectional area of the tip (True 1976, Rocker 1985, O'Loughlin et al. 2004, Chow et al. 2017).

The challenge with this approach is in correctly identifying and quantifying the magnitude of the various terms in Equation (1). A further complication is that, since the undrained shear strength (derived from the tip pressure $q_{t,FFP}$) contributes towards the shaft resistance, an iterative approach is needed in solving for the tip resistance. This may be avoided for extended base FFPs such as the free-fall sphere and STING, since the shaft resistance may be excluded from Equation (1). However, it may then be necessary to consider the additional soil mass that moves with the base (Morton et al. 2016b).

4.4.1 Hydrodynamic and soil drag

Soil drag resistance can dominate interpretation of strength in weak seabeds at shallow embedment, and accounting for this parameter is necessary for accurate strength assessments as well as in probe design. Soil drag is calculated in the same manner as for hydrodynamic drag, as

$$F_D = 0.5 C_d \rho_s A_{tip} v^2 \quad (2)$$

where ρ_s is the saturated density of the soil and C_D is the drag coefficient, the value of which depends on the penetrometer geometry. The density must be estimated (since samples are not taken) and there is uncertainty in the drag coefficient.

Consideration of the acceleration trace during the free-fall in water phase allows the fluid drag characteristics, and its variation with Reynolds number for a particular geometry, to be established (Morton et al. 2016a, O'Beirne et al. 2017). It is common to estimate the soil drag coefficient as the same as the fluid drag coefficient, which tends to vary from about 0.15 (increasing with L/D) for cylindrical penetrometers with hemispherical tips, to about 0.26 for spheres.

If the fluid and soil drag coefficient are assumed equal, the influence of the tether on drag characteristics should be considered. Depending on probe mass, geometry, fall height through water, and tether characteristics, tethered terminal velocities may be half or less than untethered velocities, as indicated by offshore data in Figure 17. Upon seabed impact the tether may tend to go slack, changing the drag resistance implied by the probe impact velocity.

4.4.2 Shaft frictional resistance

For probes that only measure acceleration or velocity, the friction along the side of the probe must be subtracted from the total resistance to

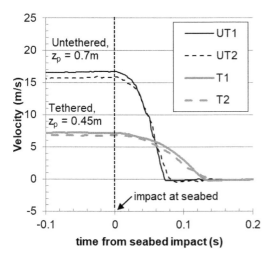

Figure 17. Influence of tethering on penetration of NAVFAC XDP drop through 30 m of water into a low plasticity clay with s_u of 10 to 20 kPa in the upper 0.75 m.

assess tip resistance. Shaft friction on the side can be estimated based on the undrained strength through an adhesion factor, α, which is commonly assumed to be the inverse of the soil sensitivity. The total shaft friction at any given penetration is:

$$Q_s = \sum \Delta A_{shaft} R_{f,shaft} \alpha s_u \qquad (3)$$

The significant uncertainty in soil sensitivity (or α), and how that varies with depth, will propagate through to the accuracy of the estimated tip resistance. The shaft friction is also found to exhibit much higher rate effects than the tip resistance (as discussed later) and hence a higher rate parameter $R_{f,shaft}$ (Steiner et al. 2014, Chow et al. 2017). The uncertainty in both the soil sensitivity and $R_{f,shaft}$ is the most significant factor for accurate estimation of soil strength using the accelerometer measurement.

4.5 Cone tip resistance and derivation of s_u

Improved estimation of soil strength can be obtained by measuring the FFP tip resistance directly with a load cell. Interpretation is similar to that for a static CPT, with the additional need to deduct drag resistance and to discount the soil strength for strain rate effects, according to:

$$s_u = \frac{q_{c,FFP} - q_D + u_2(1-\alpha_{cone}) - \sigma_{v0}}{R_{f,tip} N_{kt}}$$

$$= \frac{q_{net,d}}{R_{f,tip} N_{kt}} \sim \frac{q_{net,s}}{N_{kt}} \qquad (4)$$

where $q_{c,FFP}$ is the measured tip resistance, u_2 the pore pressure measured or estimated at the cone shoulder, α_{cone} the unequal area ratio, σ_{vo} the overburden stress, q_D the drag resistance (equivalent to F_D/A_{tip} from Equation (1)), $R_{f,tip}$ the strain rate factor for tip resistance, N_{kt} the cone factor, $q_{net,d}$ the resulting dynamic net cone resistance and $q_{net,s}$ the static net cone resistance from a conventional CPT.

4.6 Viscous rate effects

The extremely high penetration velocity of an FFP (especially at shallow depth) leads to very high strain rates in the soil that will enhance the soil strength beyond nominally undrained values. The effect of strain rate may be modelled using either a semi-logarithmic (or inverse hyperbolic sine function) or power law (including the Herschel-Bulkley formulation). The simple power law of

$$R_{f,tip} = \left(\frac{\dot{\gamma}}{\dot{\gamma}_{ref}}\right)^\beta = \left(\frac{v/D}{(v/D)_{ref}}\right)^\beta \geq 1 \qquad (5)$$

has generally been adopted as it is found to capture rate effects better than logarithmic functions over large ranges of strain rate (Biscontin & Pestana 2001). Here $\dot{\gamma}$ is the strain rate, $\dot{\gamma}_{ref}$ the reference strain rate associated with the reference value of undrained shear strength and β a strain rate parameter. The average strain rate may be linked directly to the normalised velocity, v/D. Due to the different rate effects observed for the tip and shaft resistance (see later, Figure 19) different strain rate parameters, β_{tip} and β_{shaft} are recommended for estimating $R_{f,shaft}$ and $R_{f,tip}$ in Equations (3) and (4) respectively.

The reference strain rate, and hence selection of v_{ref} and D_{ref}, should correspond to nominally undrained conditions, similar to those that would develop in a static cone penetrometer test. For instance selection of v_{ref} = 20 mm/s and D_{ref} = 35.7 mm would result in a strength that would be comparable to that measured by a 10 cm² cone penetrated at the usual 20 mm/s. For an FFP with diameter of 100 mm, travelling at velocity, v = 10 m/s, v/D = 100 s⁻¹, two orders of magnitude higher than in the static 10 cm² cone test. For this magnitude variation in strain rate, the strain rate parameter β_{tip} is typically in the range 0.03 to 0.09 for tip resistance (Lehane et al. 2009, O'Loughlin et al. 2013, 2016, Chow et al. 2017).

4.7 Example interpretation from test data

An example interpretation from centrifuge experiments in kaolin clay (Chow et al. 2017) is provided in Figure 18 to Figure 20. The interpretation based solely on acceleration measurements (but with

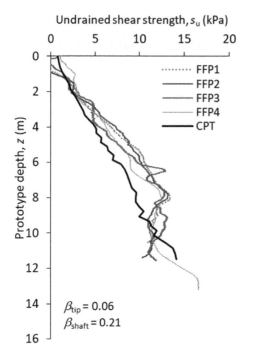

Figure 18. Estimated s_u profiles using the accelerometer method (after Chow et al. 2017).

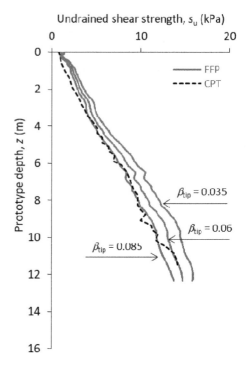

Figure 20. Deduced s_u profiles using tip load cell (after Chow et al. 2017).

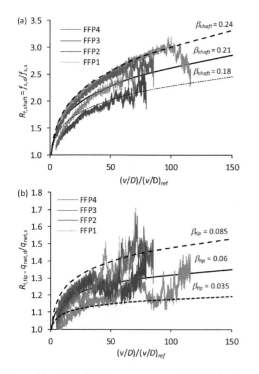

Figure 19. Back-fitted rate parameters for (a) shaft resistance and (b) net tip resistance (after Chow et al. 2017).

no account taken of any 'tethering' effects since these are negligible for the centrifuge model test) underestimates the strength deduced from static CPTs by approximately 50%, even when separate strain rate parameters (β_{tip} and β_{shaft}) are applied, as established in direct measurements from the tip and shaft load cells of the FFP (see Figure 19). This large discrepancy is caused by the uncertainty in estimating the dynamic shaft resistance reliably (Chow et al. 2017).

Figure 20 shows (for the same Chow et al. 2017 centrifuge tests) that when the interpretation is based on the tip load cell and u_2 pore pressure measurements, s_u values determined from static CPTs and FFP tests are essentially indistinguishable. The interpretation on Figure 20 uses a median $\beta_{tip} = 0.06$, with variations in β_{tip} (reducing to $\beta_{tip} = 0.035$ and increasing to $\beta_{tip} = 0.085$) leading to changes in the FFP s_u interpretation that vary by less than ±10%. Interestingly, the u_2 excess pore pressure response was mainly negative in the FFP tests and positive in the static CPTs, although the measured (negative) u_2 without adjustment was needed in Equation (4) for correcting the unequal area effect in order to provide the agreement in Figure 20.

The negative u_2 measured during the dynamic penetration has been attributed to local effects due to the cone geometry and position of the u_2 sensor (Chow et al. 2014). The phenomenon may be considered a result of Bernoulli effects, as illustrated by computational fluid dynamics analyses for free fall in water (Lucking et al. 2017, Mumtaz et al. 2018). Negative u_2 values following embedment were also found in large deformation finite element analyses for soils that show strain rate dependency of strength (Sabetamal et al. 2016). This requires further experimental validation, and is particularly important if FFPs are to be interpreted to provide Robertson-style soil behaviour types.

4.8 Penetrometer shape

Where accurate determination of s_u at shallow depths (e.g. < 1 m) is required, there may be benefits in employing 'extended base' FFPs. Examples include the plate-tipped STING (Mulhearn 2003, Fawaz et al. 2016; Chow & Airey 2014; see Figure 21a) and the free-fall sphere (Morton et al. 2016a, b; see Figure 21b). Due to the relatively large tip area of these FFPs, drag resistance becomes a significant and often dominating component of the soil resistance.

In addition to the better resolution in s_u permitted by the larger tip area, there is essentially no shaft resistance on an extended base FFP, so that the complex calculation of dynamic shaft resistance is avoided. This allows s_u to be determined indirectly from the acceleration measurements, meaning that this class of FFP does not require load cell measurements.

Example interpreted s_u profiles for the free-fall sphere are provided in Figure 22 for a lake-bed clay in Northern Ireland and a nearshore clay in the Firth of Clyde, off the West coast of Scotland. Significant penetrations were achieved, particularly in the very soft lake-bed clay, using impact velocities typically in the range 4 to 8 m/s.

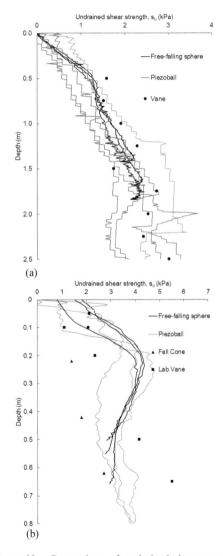

Figure 21. Extended base FFPs: (a) STING (Fawaz et al. 2016), (b) free-fall sphere.

Figure 22. Comparison of undrained shear strength profiles derived from free-fall sphere acceleration data and push-in piezoball penetration resistance: (a) a lake site (Lough Erne, Northern Ireland) and (b) a nearshore site (Firth of Clyde, Scotland) (Morton et al. 2016a).

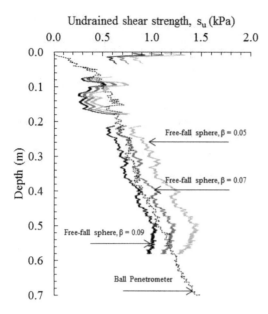

Figure 23. Sensitivity of interpreted s_u profiles to changes in the strain rate parameter (Morton et al. 2016b).

The free-fall sphere s_u profiles are seen to agree well with static piezoball tests on Figure 22, where the same piezoball factor was used (at each site) to calculate s_u from the static and FFP net resistance, and the same strain rate parameter, $\beta = 0.07$ and drag coefficient, $C_d = 0.26$ were used for both sites. Interpretation of the Clyde tests—static and free-fall—required adjustment of the bearing capacity factor at shallow depths, as discussed in the following section.

The sensitivity to uncertainty in β is demonstrated by Figure 23, which compares centrifuge s_u profiles from free-fall sphere tests with equivalent static profiles for a reconstituted carbonate clay from Laminaria (Timor Sea); β is varied by ±0.02 from the base case $\beta = 0.07$ (used for both the centrifuge and field tests). The lower bound $\beta = 0.05$ corresponds with the largest departure from the base case profile established using $\beta = 0.07$, although by less than 20%. This uncertainty is comparable to the differences that are commonly linked to uncertainty between different laboratory strength tests (Bienen et al. 2010).

4.8.1 Shallow embedment correction factors

For large diameter penetrometers, it is necessary to modify the bearing capacity factor, N_{FFP}, for shallow embedment (e.g. Rocker 1985, Aubeny & Shi 2006, White et al. 2010, Morton et al. 2016a). The bearing capacity factor tends to increase from the shallow foundation factor at the seabed interface to the deep factor at depth. In soils with a constant strength, the deep factor is typically reached within approximately four diameters (depending on the normalised strength ratio $s_u/\gamma D$). The shape of the transition to 'deep' conditions will be influenced by the strength gradient and unit weight of the soil (e.g. Stanier & White 2015).

4.9 Summary comments on FFPs

In spite of the cost-effectiveness of free-fall penetrometers (FFPs), compared with mobilising a seabed frame to conduct static penetrometer tests, the quality and usefulness of the data acquired need to be considered carefully within the context of the site investigation objectives. In the early stages of characterising the seabed, and particularly for spatially extensive surveys for pipeline routing, it may be sufficient to acquire data that merely distinguishes rock (or cemented) outcrops, from sandy deposits all the way down to fine-grained sediments of different strengths. Where soft sediments can be anticipated, quantitative strength data in the upper metre or so would be useful even at a preliminary SI stage.

Lightly instrumented FFPs are sufficient for primary categorisation of the seabed material. Penetration of the seabed will be minimal for cemented crusts, and unlikely to exceed a fraction of a metre in most types of coarse-grained (silty sands and coarser) material, regardless of the mineralogy or relative density (Stark et al. 2012). This may be sufficient during concept selection and pipeline routing studies. However, where anchors are anticipated, it may be difficult to distinguish between sandy sediments sufficiently thick for good drag anchor performance, and superficial sand layers overlying rock, resulting in much less certain anchor holding capacity.

As in other methods of site investigation, the guiding principle must be to ensure free-fall penetration (and ideally deduction of the tip resistance profile) to depths compatible with the design target. In fine-grained sediments, where quantitative assessment of the shear strength profile is important, the required accuracy needs to be considered in choosing the type of FFP.

The new generation of combined dynamic and static penetrometers, where calibration of dynamic effects can be achieved through overlapping zones of dynamic and static data (Young et al. 2011, Randolph 2016), are an excellent cost-effective approach for moderate-sized foundation or anchoring solutions. However, the deployment of such systems requires prior knowledge that the sediments are fine-grained and relatively low strength, excluding any sandy layers or other obstructions. Additionally, since calibration with respect to rate factors etc. is sensitive to the soil type, the correction may not be valid over the full depth of dynamic penetration.

For assessing the shear strength of the upper 1 m or so of the seabed, smaller FFPs, either free-falling piezocones or extended base devices, are more cost effective than larger FFPs. Ideally one or two independent static CPTs should be included for calibration, balancing the operational efficiency of FFP testing with the improved confidence and accuracy from calibration against data from static penetrometer tests (Steiner et al. 2012). Overall, though, for free-falling cone penetrometers, direct measurement of tip resistance, albeit under dynamic conditions, is essential, given the uncertainties associated with drag resistance at shallow depths and estimating the rate-enhanced shaft resistance.

5 APPLICATION OF PENETROMETER DATA IN DESIGN

5.1 *Trends of indirect and direct penetrometer correlations*

Along with the development in geotechnical site investigation, which routinely includes in situ penetrometer testing, methods to use the acquired data more directly for foundation design have become widespread. The boundary between 'indirect' and 'direct' is not clear cut, but here the former is restricted to deriving simplified profiles of fundamental soil properties such as shear strength or the equivalent (friction angle or relative density) for free-draining sediments. By contrast, direct design approaches derive foundation design parameters from the CPT data, such as modern approaches for axial pile capacity in sand as detailed in Recommended Practice 2Geo of the American Petroleum Institute (API 2011).

For shallow foundations there are a range of approaches, from directly correlating the penetrometer resistance with the foundation capacity (e.g. Lee and Randolph 2011, Pucker et al. 2013, Bienen et al. 2015), using the cone tip resistance as input in numerical analysis (e.g. Schneider et al. 2018), to inferring the soil type from the penetrometer data and using the deduced strength in traditional bearing capacity predictions (e.g. Safinus 2015). The last approach might really be considered as 'indirect' except that the method to extract the layering and strength properties is fully automated.

As with any other engineering application, though, the prediction method cannot increase the quality or relevance of the input data. Hence, in order to obtain accurate predictions, sufficient high quality penetrometer data are required at the target location in combination with a sound correlation method. Furthermore, while computational power is ever increasing, the target application may still limit what is practicable for routine design (e.g. direct input of a CPT profile into large deformation numerical analysis to obtain a jack-up spudcan load-penetration curve).

Jack-up spudcans have been the focus of a number of proposed correlation methods recently, as a result of often sparse site investigation data available for a priori predictions and relatively high proportion of incidents and accidents (Osborne et al. 2006, Hunt 2008, Jack et al. 2013). The proposed approaches to using penetrometer data in design differ in the detailed treatment of soil behaviour, but also in the computational effort required.

Assigning soil layers in the geotechnical design process retains an element of subjectivity that is difficult to remove. Automatic identification of primary geotechnical layers from in situ penetrometer data, for example following Robertson's approach of categorising the soil behaviour type index ($I_{c,RW}$ – Robertson & Wride 1998, I_B – Robertson 2016) to assign strength values, feeds into a top-down calculation to arrive at the predicted spudcan penetration resistance profile (Safinus 2015). The method offers the advantage of capturing the evolution of the soil profile (and plug) with spudcan penetration, with an example provided in Figure 24. Limitations of the approach include the reliance on correlations of questionable accuracy to infer soil unit weight and strength parameters for carbonate sediments, in particular sands.

The cone tip resistance profile can also be used directly to obtain the spudcan load-penetration curve via a correlation factor. With the bearing capacity factor N_c of a deeply penetrated foundation of approximately 9.0 and recommended average value of N_{kt} of approximately 13.5, a correlation factor of 0.67 is arrived at for clay (Osborne et al. 2011). Similar direct correlation factors for spudcans penetrating into siliceous and uncemented carbonate sand are provided in Pucker et al. (2013) and Bienen et al. (2012), respectively. This direct correlation method has been benchmarked against field data from offshore wind installations on sand sites in the North Sea (Edwards et al. 2013), with similar performance to predictions based on ISO (2012, previously SNAME) – see Figure 25.

Layered soil profiles introduce the issue of scaling, as a large size object such as a spudcan will have an averaging effect, while a penetrometer will resolve finer detail of strength variation. In the guidelines originating from the InSafe JIP (Osborne et al. 2011) a strength averaging approach is adopted to account for the apparent size effect.

Two other aspects of scale are important. The first involves differences in drainage regime for the

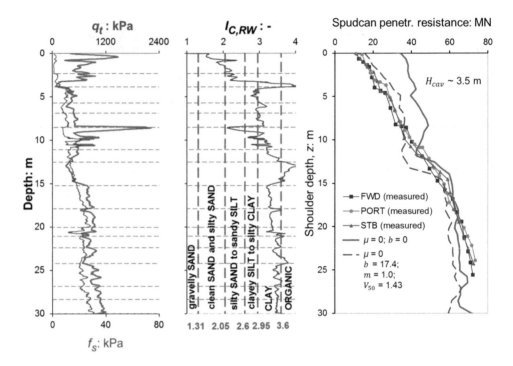

Figure 24. Spudcan load-penetration curve prediction (after Safinus 2015).

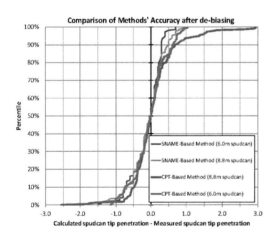

Figure 25. Comparison of the performance of direct CPT correlation and predictions based on ISO (2012).

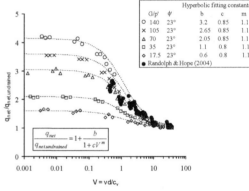

Figure 26. Normalised enhancement of penetration resistance with non-dimensional velocity.

penetrating object compared with a penetrometer. The second concerns the different modes of failure arising in strongly stratified profiles, i.e. with sand layers interleaved with layers of fine grained sediments.

The drainage regime may be close to fully undrained for a large diameter foundation such as a penetrating spudcan, even in sediments that are essentially free-draining during a penetrometer test. These differences in drainage characteristics lead to variations in the mobilised resistance, as illustrated in Figure 4 and Figure 5. The guidelines resulting from the InSafe JIP (Osborne et al. 2011) recommend modifying the factor of N_c/N_{kt} by the ratio between spudcan and penetrometer 'q_{net}' values arising from partial drainage, as illustrated in Figure 26. Values of q_{net} are estimated from the consolidation curves for normalised velocities

$V = vD/c_v$ using the spudcan and penetrometer diameters.

Erbrich (2005), Amodio et al. (2015) and Erbrich et al. (2015) provide detailed discussion of the effect of drainage on spudcan response, including additional consideration of cyclic degradation of the carbonate soils prevalent in areas of oil and gas developments offshore Australia (North West Shelf, Bass Strait).

With respect to the different penetration mechanisms that occur in stratified profiles, some methods (e.g. Safinus 2015) take specific account of effects such as squeezing of soft material overlying stronger material, punch-through for 'sand over clay' profiles, and allowance for a plug of stronger soil being trapped ahead of the penetrating spudcan. Simpler approaches are based on direct correlation to obtain an estimate of the spudcan load-penetration curve from the cone tip resistance, including consideration of the averaging distance. A method of this type is described by Bienen et al. (2015) for stratified 'sand over clay' profiles that may bear the risk of punch-through failure.

Jack-up spudcans are only one application example, of course, with other relevant applications affected by soil stratification including laterally loaded piles and anchor response. The example provided in Schneider et al. (2018) illustrates the merit of using in situ penetrometer data as input into numerical analysis to obtain foundation capacity in soil with layering. Another example, but for suction caisson capacity, is given below.

This section has illustrated a range of approaches to utilise penetrometer data in design. Penetrometer based methods need to be evaluated alongside more traditional design approaches, using data from actual foundation performance, in order to improve confidence in the methods and encourage further evolution. This also offers the opportunity of evidence-based selection of the most economical prediction method, balancing an appropriate level of sophistication relative to the available input data.

5.2 Direct and indirect application of penetrometer data in numerical analysis

Most foundation and anchor design is based on 'interpreted' penetrometer data, where simplified piecewise linear variations of shear strength are fitted to profiles of net penetrometer tip resistance using appropriate bearing (i.e. cone, T-bar, ball, etc.) factors. Schneider et al. (2018) discuss the potential errors that may arise from this approach, comparing results obtained using either an idealised strength profile, or the actual variation of (factored) net cone resistance.

Figure 27 shows the strength profile deduced from CPT data using a cone factor of $N_{kt} = 13.5$, from the second example (C6) considered in Schneider et al. (2018). The CPT profile is compared with the corresponding idealised s_u profile of $s_u = \max(1.9, 1.43z)$ kPa, where z is the depth in m. Corresponding ratios of the idealised and 'direct' (i.e. from the CPT profile) shear strengths are shown in Figure 28.

The idealised strength profile shown in Figure 27 might not be considered an appropriately 'conservative' idealisation for use in design. However, in practice such profiles are generally fitted (conservatively) through a number of CPT profiles, rather than being a strict lower bound to all profiles. The purpose here is to explore what potential over prediction of capacity might arise from the mismatch of the shear strength profiles in the upper few metres.

For the shallow foundation systems considered by Schneider et al. (2018), the vertical capacity of strip foundations of widths 5 to 20 m estimated using the CPT data directly were some 15% lower than those based on the idealised s_u profile. More significant differences were found for a foundation system compromising two linked skirted foundations, with skirt tips embedded 1.75 m, where the moment capacity using the CPT data was less than 50% of that from the idealised profile.

Figure 27. Strength profiles from Schneider et al. (2018).

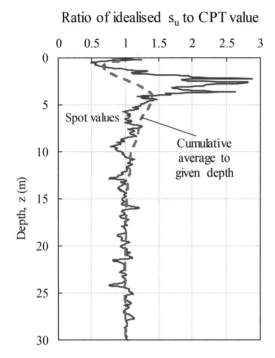

Figure 28. Ratios of spot values and cumulative averages for the idealised and 'direct' s_u profiles.

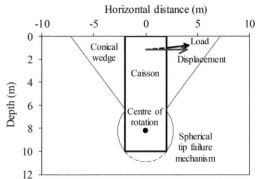

Figure 29. Suction anchor and upper bound failure mechanism.

Table 1. Strength and anchor capacities for different s_u profiles.

	CPT data	Idealised s_u	Ratio
Average s_u (0 to 10 m)	6.7 kPa	7.3 kPa	0.92
Anchor capacity (rotating)	828 kN	957 kN	0.86
Anchor capacity (translating)	3041 kN	3278 kN	0.93

Considering the profiles of cumulative average strength ratio shown in Figure 28, it might be expected that the capacity of a suction anchor would be relatively insensitive to the detailed stratigraphy observed from the CPT data, in particular the weak zone between 2 and 5 m depth. However, there are still significant differences.

Common applications for suction anchors for deep water developments include relatively small caissons for manifolds and pipeline end terminations. Typically, lateral loading for such caissons is applied near the seabed surface, leading to a rotational failure mode.

Figure 29 shows a 4 m diameter caisson embedded to a depth of 10 m, with loading applied at a depth of 1 m at an angle of 5° from the horizontal. The corresponding upper bound failure mechanism (Murff & Hamilton 1993) is as indicated, with a conical wedge failure extending to a depth of about 6 m, below which there is a rotating spherical failure.

Upper bound capacities evaluated using either the factored CPT data directly or the linearised s_u profile shown in Figure 27 are 828 kN and 957 kN respectively. The nearly 15% difference may be attributed partly to the approximately 9% difference in the cumulative average shear strength for the two profiles between 0 and 10 m. However, the weaker upper soil layer leads to further difference in capacity for a caisson with loading applied near the ground surface. By contrast, capacities for a purely horizontally translating caisson differ by just the difference in average shear strength over the relevant depth range. These results are summarised in Table 1.

It is possible to obtain similar capacities by judicious choice of the idealised s_u profile (for example, a crustal shear strength of 2.1 kPa, below which the shear strength is given by $s_u = -4.2 + 2z$ kPa). However, offshore design tends to follow a phased approach, with a geotechnical interpretive report providing simplified strength profiles for a range of possible foundation or anchoring systems, and detailed design undertaken in a subsequent engineering phase. The potential for significant overestimation of capacities due to subtle limitations in the idealised strength profiles is evident.

6 SUMMARY

This paper has considered a number of issues pertaining to penetrometer testing and interpretation with respect to offshore geotechnical design. The important trends are for improved cost-effectiveness, timely acquisition of quantitative strength (or other design) data, and increased sophistication in testing and control procedures.

In keeping with these trends, the paper has included extended discussion of free-fall penetrometers and the techniques required to obtain equivalent 'static' strength or resistance parameters. Similarly, the potential for integrating box core sampling and deck-based penetrometer testing in early phases of (geophysical focused) site investigation has been discussed. The manner of the box-core tests, with close control of load and displacement on different axes of motion, is extendable to seabed-based testing, consistent with the underlying objective of 'bringing the laboratory to the seabed'.

A final consideration has been the manner in which penetrometer data should be utilised in design calculations. Important issues here are (a) to make appropriate allowances for the differences in scale between penetrometers, seabed layering and the design target; and (b) the increasing potential to incorporate the full detail of CPT data in design calculations.

ACKNOWLEDGEMENTS

This work forms part of the activities of the Centre for Offshore Foundation Systems at UWA, currently supported as a node of the Australian Research Council Centre of Excellence for Geotechnical Science and Engineering. The authors are grateful for support from the RIGSS JIP sponsors: Fugro, Shell, Total and Woodside Energy. In addition, Mark Randolph acknowledges support through the Fugro Chair in Geotechnics and David White acknowledges support through the Shell Chair in Offshore Engineering.

REFERENCES

Amodio, A., Erbrich, C.T., Murugavel, V. & Moyle, I. 2015. Re-visiting Yolla—managing jack-up storm stability: geotechnical assessment. *Proc. 15th Int. Conf. The Jack-Up Platform Design, Construction & Operation*, London.

API 2011. *RP 2GEO—Geotechnical and foundation design considerations*. American Petroleum Institute, Washington.

Aubeny, C.P. & Shi, H. 2006. Interpretation of impact penetration measurements in soft clays. *J. Geotech. & Geoenv. Eng.*, ASCE, 132(6): 770–777.

Beard, R.M., 1985. Expendable Doppler penetrometer for deep ocean sediment measurements. *ASTM Special Technical Publication*, 883: 101–124.

Been, K., Jefferies, M.G., & Hachey, J. 1991. The critical state of sands. *Géotechnique*, 41(3): 365–381.

Bienen, B., Cassidy, M.J., Randolph, M.F. & Teh, K.L. 2010. Characterisation of undrained shear strength using statistical methods. *Proc. 2nd Int. Symp. Frontiers in Offshore Geotech. ISFOG2010*, Taylor & Francis, London: 661–666.

Bienen, B., Pucker, T. & Henke, S. 2012. Cone penetrometer based spudcan penetration prediction in uncemented carbonate sand. *Proc. Offshore Technology Conf.*, Houston, Paper OTC 23002.

Bienen, B., Qiu, G. & Pucker, T. 2015. CPT correlation developed from numerical analysis to predict jack-up foundation penetration into sand overlying clay. *Ocean Engineering*, 108: 216–226.

Biscontin, G. & Pestana, J.M. 2001. Influence of peripheral velocity on vane shear strength of an artificial clay. *Geotechnical Testing Journal*, 24(4): 423–429.

Blake, A., O'Loughlin, C.D., Morton, J., O'Beirne, C., Gaudin, C. & White, D.J. 2016. In-situ measurement of the dynamic penetration of free-fall projectiles in soft soils using a low cost inertial measurement unit. *Geotechnical Testing Journal*, 39(2): 235–251.

Bolton, M.D., 1986. The strength and dilatancy of sands. *Géotechnique*, 36(1): 65–78.

Boscardin, A.G. & DeGroot, D.J. 2015. Evaluation of a toroid for model pipeline testing of very soft offshore box core samples. *Proc. 3rd Int. Symp. Frontiers in Offshore Geotech. ISFOG2015*, Taylor & Francis, London: 363–368.

Chow, S.H. & Airey, D.W. 2014. Free-falling penetrometers: a laboratory investigation in clay. *J. Geotech. & Geoenv. Eng.*, ASCE, 140(1): 201–214.

Chow, S.H., Bienen, B. & Randolph, M.F. 2018. Rapid penetration of piezocones in sand. *Proc. Int. Symp. Cone Penetration Testing, CPT'18*, Delft.

Chow, S.H., O'Loughlin, C.D. & Randolph, M.F. 2014. Soil strength estimation and pore pressure dissipation for free-fall piezocone in soft clay. *Géotechnique*, 64(10): 817–827.

Chow, S.H., O'Loughlin, C.D., White, D.J. & Randolph, M.F. 2017. An extended interpretation of the free-fall piezocone test in clay. *Géotechnique*, 67(12): 1090–1103.

Clukey, E.C., Aubeny, C., Zakeri, A., Randolph, M.F., Sharma, P.P., White, D.J., Sancio, R. & Cerkovnik, M. 2017. A perspective on the state of knowledge regarding soil-pipe interaction for SCR fatigue assessments. *Proc. Offshore Technology Conf.*, Houston, Paper OTC 27564.

DNVGL 2017. *Pipe-soil interaction for submarine pipelines. Recommended Practice F-114*, DNVGL, Oslo.

Edwards, D., Bienen, B., Pucker, T. & Henke, S. 2013. Evaluation of the performance of a CPT-based correlation to predict spudcan penetrations using field data. *Proc. 14th Int. Conf. The Jack-Up Platform—Design, Design, Construction & Operation*, London.

Erbrich, C.T. 2005. Australian frontiers—spudcans on the edge. *Proc. Int. Symp. on Frontiers in Offshore Geotechnics, ISFOG*, Perth: 49–74.

Erbrich, C.T., Amodio, A., Krisdani, H., Lam S.Y., Xu, X. & Tho, K.K. 2015. Re-visiting Yolla—new insights on spudcan penetration. *Proc. 15th Int. Conf. The Jack-Up Platform Design, Construction & Operation*, London.

Fawaz, A., Teoh, A., Airey, D.W. & Hubble, T. 2016. Soil strength in the Murray River determined from a free falling penetrometer. Proc. *5th Int. Conf. on Geo-*

tech. & Geophysical Site Characterisation, ISC'5 Gold Coast: 1217–1222.

Furlong, A., Osler, J.C., Christian, H., Cunningham, D. & Pecknold, S. 2006. The moving vessel profiler (MVP) - a rapid environmental assessment tool for the collection of water column profiles and sediment classification. Proc. Undersea Defence Technology Pacific Conf., San Diego: 1–13.

Geirnaert, K., Staelens, P., Deprez, S., Noordijk, A. & Van Hassent, A. 2013. Innovative free fall sediment profiler for preparing and evaluating dredging works and determining the nautical depth. WODCON XX: 1–11.

Hunt, R., 2008. Achieving operational excellence for jack-up rig deployments—how shall we get there? Proc. 2nd Jack-Up Asia Conf. and Exhibition, Singapore.

ISO, 2012. Petroleum and natural gas industries—site-specific assessment of mobile offshore units—part 1: Jack-ups, 19905–1, Int. Organization for Standardization, Geneva.

Jack, R.L., Hoyle, M.J.R., Smith, N.P. & Hunt, R.J. 2013. Jack-up accident statistics—a further update. Proc. 14th Int. Conf. The Jack-Up Platform—Design, Const. & Oper., London.

Kelleher, P., Low, H.E., Jones, C., Lunne, T., Strandvik, S. & Tjelta, T.I. 2011. Strength measurement in very soft upper seabed sediments. Proc. 2nd Int. Symp. Frontiers in Offshore Geotech. ISFOG2010, Taylor & Francis, London: 283–288.

Kopf, A., Stegmann, S., Krastel, S., Foerster, A., Strasser, M. & Irving, M. 2007. Marine deep-water free-fall CPT measurements for landslide characterisation off Crete, Greece (Eastern Mediterranean Sea), part 2: initial data from the western Cretan Sea. Proc. Int. Conf. Submarine Mass Movements and Their Consequences. Springer, Netherlands: 199–208.

Lee, J. & Randolph, M.F. 2011. Penetrometer-based assessment of spudcan penetration resistance. J. Geotech. & Geoenv. Eng., ASCE, 137(6): 587–596.

Lehane, B.M., O'Loughlin, C.D., Gaudin, C., & Randolph, M.F. 2009. Rate effects on penetrometer resistance in kaolin. Géotechnique, 59(1), 41–52.

Low, H.E., Randolph, M.F., Rutherford, C.J., Bernard, B.B. & Brooks, J.M. 2008. Characterization of near seabed surface sediment. Proc. Offshore Technology Conf., Houston, Paper OTC 19149.

Lucking, G., Stark, N., Lippmann, T. & Smyth, S. 2017. Variability of in situ sediment strength and pore pressure behavior of tidal estuary surface sediments, Geo-Marine Letters, 37(5): 441–456.

Mahmoodzadeh, H., Wang, D. & Randolph, M.F. 2015. Interpretation of piezoball dissipation testing in clay. Géotechnique, 65(10): 831–842.

Morton, J.P., O'Loughlin, C.D. & White, D.J. 2016a. Estimation of soil strength in fine-grained soils by instrumented free-fall sphere tests. Géotechnique, 66(12): 959–968.

Morton, J.P., O'Loughlin, C.D. & White, D.J. 2016b. Centrifuge modelling of an instrumented free-fall sphere for measurement of undrained strength in fine-grained soils. Canadian Geotech. J., 53: 918–929.

Mosher, D.C., Christian, H., Cunningham, D., Mackillop, K., Furlong, A., & Jarrett, K. 2007. The Harpoon Free Fall Cone Penetrometer for Rapid Offshore Geotechnical Assessment. Proc. 6th Int. Offshore Site Investigation and Geotechnics Conf., London: 195–202.

Mulhearn, P.J. 2003. Influences of penetrometer tip geometry on bearing strength estimates. Int. J. of Offshore and Polar Eng., 13(1): 73–78.

Mumtaz, M.B., Stark, N. & Brizzolara, S. 2018. Pore pressure measurements using a portable free fall penetrometer. Proc. Int. Symp. Cone Penetration Testing, CPT'18, Delft.

Murff, J.D. & Hamilton, J.M. 1993. P-ultimate for undrained analysis of laterally loaded piles, J. Geotechnical Eng., ASCE, 119 (1): 91–107.

O'Beirne, C., O'Loughlin, C.D. & Gaudin, C. 2017. A release-to-rest model for dynamically installed anchors, J. Geotech. & Geoenv. Eng., ASCE, 143(9): 04017052.

O'Loughlin, C.D., Blake, A.P. & Gaudin, C. 2016. Towards a design method for dynamically embedded plate anchors, Géotechnique, 66(9): 741–753.

O'Loughlin, C.D., Gaudin, C., Morton, J.P. & White, D.J. 2014. MEMS accelerometers for measuring dynamic penetration events in geotechnical centrifuge tests. Int. J. Physical Modelling in Geotech., 14(2): 31–39.

O'Loughlin, C.D., Randolph, M.F. & Richardson, M.D. 2004. Experimental and theoretical studies of deep penetrating anchors. Proc. Offshore Tech. Conf., Houston, OTC 16841.

O'Loughlin, C.D., Richardson, M.D., Randolph, M.F. & Gaudin, C. 2013. Penetration of dynamically installed anchors in clay. Géotechnique, 63(11): 909–919.

Orenberg, P., True, D.G., Bowman, L., Herrmann, H. & March, R. 1996. Use of a dropped dynamic penetrometer in cohesionless soil. Proc. Offshore Tech. Conf., Houston, Paper OTC 8027.

Osborne, J.J., Pelley, D., Nelson, C. & Hunt, R. 2006. Unpredicted jack-up foundation performance. Proc. Jack-Up Asia Conf. and Exhibition, Singapore.

Osborne, J.J., Teh, K.L., Houlsby, G.T., Cassidy, M.J., Bienen, B. & Leung, C.F. 2011. 'InSafeJIP' – Improved Guidelines for the Prediction of Geotechnical Performance of Spudcan Foundations During Installation and Removal of Jack-up Units, Final Guidelines of the InSafe Joint Industry Project Report Number EOG0574-Rev1. RPSEnergy, Woking, UK.

Peuchen, J., Looijen, P.N. & Stark, N. 2017. Offshore characterisation of extremely soft sediments by free fall penetrometer. Proc. 8th Int. Conf. Offshore Site Investigation and Geotechnics, Soc. for Underwater Tech., London, 1: 370–377.

Peuchen, J. & Westgate, Z. 2018. Defining geotechnical parameters for surface-laid subsea pipe-soil interaction. Proc. Int. Symp. Cone Penetration Testing, CPT'18, Delft.

Pucker, T., Bienen, B. & Henke, S. 2013. CPT based prediction of foundation penetration in siliceous sand. Applied Ocean Research, 41: 9–18.

Randolph, M.F. 2016. New tools and directions in offshore site investigation—ISC'5 Keynote Lecture. Australian Geomechanics, 51(4): 81–92.

Randolph, M.F., Low, H.E. & Zhou, H. 2007. In situ testing for design of pipeline and anchoring systems. Keynote paper, Proc. 6th Int. Conf. Offshore Site Investigation and Geotechnics, Society for Underwater Technology, London: 251–262.

Robertson, P.K. 2016. Cone penetration test (CPT)-based soil behaviour type (SBT) classification system—an update. *Canadian Geotech. J.*, 53(12): 1910–1927.

Robertson, P.K. & Wride, C.E. 1998. Evaluating cyclic liquefaction potential using the cone penetration test. *Canadian Geotech. J.*, 35(3): 442–459.

Rocker, K. 1985. *Handbook for marine geotechnical engineering*. Naval Civil Eng. Lab., Port Hueneme, CA.

Sabetamal, H., Carter, J.P., Nazem, M. & Sloan, S.W. 2016. Coupled analysis of dynamically penetrating anchors. *Computers & Geotechnics*, 77: 26–44.

Safinus, S. 2015. *Estimation of spudcan penetration resistance in stratified soils from field piezocone penetrometer data*. PhD Thesis, University of Western Australia.

Schneider, J.A., Doherty, J.P., Randolph, M.F. & Krabbenhoft, K. 2018. Direct use of CPT data for numerical analysis of VHM loading of shallow foundations. *Proc. Int. Symp. Cone Penetration Testing, CPT'18*, Delft.

Stanier, S. & White D.J. 2015. Shallow penetrometer penetration resistance. *J. Geotech. & Geoenv. Eng.*, ASCE, 141(3): 04014117.

Stark, N., Hanff, H. & Kopf, A.J. 2009. Nimrod: a tool for rapid geotechnical characterization of surface sediments. *Sea Technology*, 50(4): 10–14.

Stark, N., Hay, A.E. & Trowse, G. 2014. Cost-effective geotechnical and sedimentological early site assessment for ocean renewable energies. *2014 Oceans—St. John's*: 1–8.

Stark, N., Wilkens, R., Ernsten, V.B., Lambers-Huesman, M., Stegmann, S. & Kopf, A.J. 2012. Geotechnical properties of sandy seafloor and the consequences for dynamic penetrometer interpretations: quartz sand versus carbonate sand. *Geotechnical and Geological Engineering*, 30: 1–14.

Stegmann, S., Morz, T. & Kopf, A.J. 2006. Initial results of a new free fall-cone penetrometer (FF-CPT) for geotechnical in situ characterisation of soft marine sediments. *Norsk Geologisk Tidsskrift*, 86(3): 199–208.

Steiner, A., L'Heureux, J.S., Kopf, A., Vanneste, M., Longva, O., Lange, M. & Haflidason, H. 2012. An in-situ free-fall piezocone penetrometer for characterizing soft and sensitive clays at Finneidfjord (Northern Norway). *Proc. Int. Conf. Submarine Mass Movements and Their Consequences*. Adv. Natural & Tech. Hazards, 31, Springer: 99–109.

Steiner, A., Kopf, A.J., L'Heureux, J.S., Kreiter, S., Stegmann, S., Haflidason, H. & Moerz, T. 2014. In situ dynamic piezocone penetrometer tests in natural clayey soils–a reappraisal of strain-rate corrections. *Can. Geotech. J.*, 51(3): 272–288.

Stephan, S., Kaul, N. & Stark, N. 2011. LIRmeter: A new tool for rapid assessment of sea floor parameters. Bridging the gap between free-fall instruments and frame-based CPT. *Proc. MTS/IEEE OCEANS 2011*. IEEE *OCEANS*.

Stoll, R.D. & Akal, T. 1999. XBP-Tool for rapid assessment of seabed sediment properties. *Sea Technology*, 40(2): 47–51.

Thompson, D., March, R. & Herrmann, H. 2002. Groundtruth results for dynamic penetrometers in cohesive soils. In *Proceedings of OCEANS 2002, MTS/IEEE—Marine Frontiers: Reflection of the Past, Visions of the Future*, (4): 2117–2123.

True, D.G. 1976. *Undrained vertical penetration into ocean bottom soils*. PhD thesis, University of California, Berkeley.

White, D.J., Gaudin, C., Boylan, N. & Zhou, H. 2010. Interpretation of T-bar penetrometer tests at shallow embedment and in very soft soils. *Canadian Geotech. J.*, 47(2): 218–229.

White, D.J. & Randolph, M.F. 2007. Seabed characterisation and models for pipeline-soil interaction. *Int. J. Offshore & Polar Engng.*, 17(3): 193–204.

White, D.J., O'Loughlin, C.D., Stark, N. & Chow, S.H. 2018. Free fall penetrometer tests in sand: Determining the equivalent drained resistance. *Proc. Int. Symp. Cone Penetration Testing, CPT'18*, Delft.

White, D.J., Stanier, S.A., Schneider, M.A., O'Loughlin, C.D., Chow, S.H., Randolph, M.F., Draper, S.D. Mohr, H, Morton, J.P., Peuchen, J., Fearon, R., Roux, A. & Chow, F.C. 2017. Remote intelligent geotechnical seabed surveys—technology emerging from the RIGSS JIP. *Proc. 8th Int. Conf. Offshore Site Investigation and Geotechnics*, Society for Underwater Technology, London, 2: 1214–1222.

White, D.J., Clukey, E.C., Randolph, M.F., Boylan, N.P., Bransby, M.F., Zakeri, A., Hill, A.J. & Jaeck, C. 2017. The state of knowledge of pipe-soil interaction for on-bottom pipeline design. *Proc. Offshore Tech. Conf.*, Houston, Paper OTC 27623.

Yan, Y., White D.J. & Randolph M.F. 2014. Cyclic consolidation and axial friction on seabed pipelines. *Géotechnique Letters*, 4: 165–169.

Yan Y., White D.J. & Randolph M.F. 2017. Elastoplastic consolidation solutions for scaling from shallow penetrometers to pipelines. *Canadian Geotech. J.*, 54: 881–895.

Yan Y. White D.J. & Randolph M.F. 2011. Penetration resistance and stiffness factors in uniform clay for hemispherical and toroidal penetrometers *Int. J. of Geomech.*, ASCE, 11: 263–275.

Young, A.G., Bernard, B.B., Remmes, B.D., Babb, L.V. & Brooks, J.M. 2011. "CPT Stinger"—an innovative method to obtain cpt data for integrated geoscience studies. *Proc. Offshore Tech. Conf.*, Houston, Paper OTC 21569.

Yuan, F., White, D.J. & O'Loughlin, C.D. 2017. The evolution of seabed stiffness during cyclic movement in a riser touchdown zone on soft clay. *Géotechnique*, 67(2): 127–137.

Inverse filtering procedure to correct cone penetration data for thin-layer and transition effects

R.W. Boulanger & J.T. DeJong
Department of Civil and Environmental Engineering, University of California, Davis, USA

ABSTRACT: This paper presents an inverse filtering procedure for developing estimates of "true" cone penetration tip resistance and sleeve friction values from measured cone penetration test data in interlayered soil profiles. Results of prior studies of cone penetration in layered soil profiles are utilized for developing and evaluating the inverse filtering procedure. The inverse filtering procedure has three primary components: (1) a model for how the cone penetrometer acts as a low-pass spatial filter in sampling the true distribution of soil resistance versus depth, (2) a solution procedure for iteratively determining an estimate of the true cone penetration resistance profile from the measured profile given the cone penetration filter model, and (3) a procedure for identifying sharp transition interfaces and correcting the data at those interfaces. The details of the inverse filtering procedure presented herein were developed with a focus on liquefaction problems, but the concepts and framework should be applicable to other problems. Example applications of the inverse filtering procedure are presented for four CPT soundings illustrative of a range of soil profile characteristics. The proposed procedure provides an objective, repeatable, and automatable means for correcting cone penetration test data for thin-layer and transition zone effects.

1 INTRODUCTION

The cone penetration test (CPT) provides excellent stratigraphic detail and information for estimating a wide range of soil properties, but the spatial resolution of cone tip resistance (q_t) and sleeve friction (f_s) measurements is still limited by the physical volume of soil around a cone tip that influences those measurements. Measurements of q_t are most strongly influenced by soils within about 10–30 cone diameters (d_c) of the cone tip, which corresponds to influence zones of 0.35–1.3 m thickness for standard 10 cm^2 and 15 cm^2 cones. Measurements of q_t and f_s therefore depend on the sequence and properties of all soils within the zone of influence, such that the cone acts as a low-pass spatial filter on the true distribution of soil resistance in a soil profile. This physical low-pass spatial filtering removes information at the shorter physical wavelengths [m], corresponding to higher spatial frequencies [cycles/m], that are necessary for defining sharp interfaces between soils with different properties. The resulting spatial smoothing of information at interfaces in interbedded soil deposits is well recognized in practice (e.g., Lunne et al. 1997, Mayne 2007) and has been the focus of considerable study.

The loss of detail at shorter physical wavelengths during cone penetration in layered soil profiles may not be of importance in some areas of practice, but there are certain situations where the resulting "thin layer" and "transition zone" effects can be sufficiently important to warrant evaluating. For example, thin layer effects can be important for liquefaction methodologies, depending on the analysis procedures, soil conditions, and seismic loading (as discussed in Boulanger et al. 2016). The use of simplified one-dimensional (1D) liquefaction vulnerability indices (LVIs) can overestimate the potential for liquefaction induced deformations if the predicted intervals of liquefaction triggering are primarily associated with numerous thin layers or transition zones. In other cases, the results of 1D-LVI's may be insensitive to thin layer and transition zones if those zones are a small portion of the predicted intervals of liquefaction triggering. For nonlinear dynamic analyses (NDAs) of sites with interbedded soils, the representative properties assigned to the liquefiable interlayers can similarly benefit from accounting for thin layer and transition zone effects in some situations and be relatively unaffected in others. More commonly, thin layer and transition zone effects are just one factor among several that can contribute to an accumulation of conservatism or bias in predicted behaviors (e.g., Boulanger et al. 2016, Munter et al. 2017, Cox et al. 2017).

This paper presents an inverse filtering procedure for developing estimates of "true" cone penetration tip resistance and sleeve friction values from measured cone penetration test data. Results of prior studies are briefly reviewed and the available theoretical and experimental data for thin layer and transition zone effects in idealized two—and three-layer soil profiles are utilized for developing the inverse filtering procedure. The inverse filtering procedure and each of its three primary components are described, including: (1) the model for how the cone penetrometer acts as a low-pass spatial filter in sampling the true distribution of soil resistance versus depth, (2) the solution procedure for iteratively determining an estimate of the true cone penetration resistance profile from the measured profile given the cone penetration filter model, and (3) the procedure for identifying sharp transition interfaces and correcting data at those interfaces. Example applications of the inverse filtering procedure are then presented for four CPT soundings illustrative of a range of soil profile characteristics. The proposed procedure, which is easy to automate and perform, is shown to work well for a range of stratigraphies. It is hoped that future experience with application of the procedure in practice will lead to further improvements.

2 PAST STUDIES OF PENETRATION IN LAYERED SOIL PROFILES

Thin layer and transition zone effects have been studied extensively, including contributions from the authors listed in Table 1. These studies have utilized elastic analyses, nonlinear analyses (cavity expansion and axisymmetric models), and physical measurements (1 g physical models, centrifuge models, field data). In all cases except the elastic analyses by Yue and Yin (1999), these studies have focused on idealized profiles with two or three uniform soil layers in different sequences; e.g., a stronger soil over a weaker soil, a weaker soil over a stronger soil, a stronger soil layer embedded in weaker soil, or a weaker soil layer embedded in a stronger soil.

The schematic in Figure 1 shows the case of a sand layer embedded in a clay deposit to illustrate both the thin layer and transition zone effects. The measured tip resistance (denoted as q^m) will smoothly increase as the cone approaches and enters the stronger layer and then smoothly decrease as the cone approaches and then enters the underlying weaker soil. The "true" tip resistance (denoted as q^t) is the value that would have been measured in this same soil if the measurement was free of the influence of the overlying and underlying weaker clay soils. The "transition" zones are those intervals near the layer interfaces over which q^m smoothly increases or decreases even though q^t abruptly changes. The thin layer effect occurs when the peak q^m becomes smaller than the corresponding q^t, with the error increasing as the stronger layer's thickness decreases. The thin layer factor (K_H), defined as q^t divided by the peak q^m for the layer, therefore increases as layer thickness decreases.

Table 1. Past studies of cone penetration in layered soil profiles.

Authors	Primary focus
Elastic analyses:	
Sayed & Hamed (1987)	Spherical and cylindrical cavity expansion in layered elastic system
Vreugdenhil et al. (1994)	Elastic solutions for stress distributions in layered elastic system
Yue & Yin (1999)	Elastic solutions for stresses in a multi-layered system
Nonlinear analyses:	
Van den Berg et al. (1996)	Axisymmetric penetration analysis in layered sand and clay
Ahmadi & Robertson (2005)	Axisymmetric penetration analysis in layered sand and clay
Xu & Lehane (2008)	Spherical cavity expansion analogue for layered sand and clay
Walker & Yu (2010)	Axisymmetric penetration analysis in layered clay
Mo et al. (2017)	Cavity expansion analysis for layered sand and clay.
Physical data:	
Treadwell (1976)	Chamber tests of cone penetration in layered sand
Meyerhof & Valsangka (1977)	Model tests of piles and cones in layered sand and clay
Foray & Pautre (1988)	Chamber tests of cone penetration in layered sand
Canou (1989)	Chamber tests of cone penetration in layered sand
Youd et al. (2001)	Interpretation of field data
Hird et al. (2003)	Chamber tests of piezocones for thin sand/silt layers in clay
Silva & Bolton (2004)	Centrifuge tests of cone penetration in layered sand
Mlynarek et al. (2012)	Chamber tests of cone penetration in layered sand and clay
Mo et al. (2015)	Centrifuge tests of cone penetration in layered sand
Tehrani et al. (2018)	Chamber tests of cone penetration in layered sand

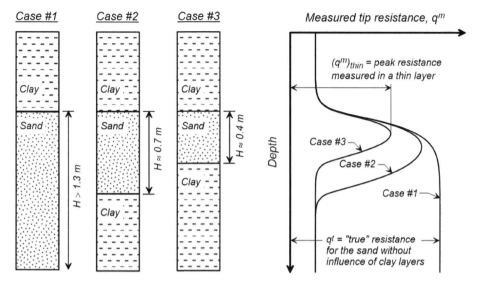

Figure 1. Schematic of thin layer effect for a sand layer embedded in a clay layer (modified from Robertson and Fear 1995).

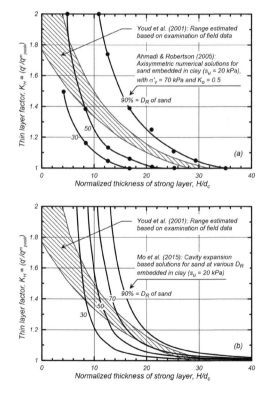

Figure 2. Thin layer factors inferred from field data (Youd et al. 2001) compared to: (a) axisymmetric numerical analyses by Ahmadi and Robertson (2005), and (b) cavity expansion based solutions by Mo et al. (2017).

Thin layer factors from three different studies are plotted versus normalized layer thickness (H/d_c) in Figures 2a and 2b to illustrate the range of findings in the literature. The range presented by Youd et al. (2001), and shown in both figures for reference purposes, was inferred from examinations of field data, although the details of the field data and interpretation method were not described. The results shown for Ahmadi and Robertson (2005) in Figure 2a are from axisymmetric nonlinear analyses of a sand layer with relative densities (D_R) of 30, 50 and 90% with $\sigma'_v = 70$ kPa and $K_o = 0.5$ (producing q^t of about 3.2, 5.8, and 22 MPa, respectively) embedded in soft clay with an undrained shear strength (s_u) of 20 kPa (producing a q^t of about 0.9 MPa). The corresponding thin layer factors increased with the D_R of the sand layer, which alternatively can be identified as having q^t_{strong}/q^t_{weak} ratios of about 3.6, 6.4, and 24, respectively. Ahmadi and Robertson (2005) also showed their thin layer factors to decrease with increasing effective confining stress, which is consistent with the dependence of sand dilation angle on confining stress used in their analyses. The results shown for Mo et al. (2017) in Figure 2b are based on cavity expansion analyses for a sand layer with D_R of 30, 50, 70, and 90% embedded in clay ($s_u = 20$ kPa), with the properties chosen to be equivalent to those used by Ahmadi and Robertson (2005). The corresponding thin layer factors are similar to those of Ahmadi and Robertson (2005), but become steeper and exceed values of 2.0 at larger values of the normalized layer thickness. The q^t_{strong}/q^t_{weak} ratios for the cases analyzed by Mo et al. (2017) were not reported, but should be similar to those for the cases analyzed by Ahmadi and Robertson (2005). The results of these and other studies for idealized three-layer profiles produce K_H values that are less than about 1.1 when the strong layer thickness is greater than about 30 cone diameters, increase as the q^t_{strong}/q^t_{weak} ratio increases, and can exceed values of 2.0 for layers that are less than about 10 cone diameters thick if the q^t_{strong}/q^t_{weak} ratio is large enough.

Transition zones in idealized two layer profiles can also be described in terms of the sensing and development distances. When the cone tip is in the upper layer, the sensing distance is defined as the greatest distance between the cone tip and the top of the underlying layer for which q^m in the upper layer is "affected" by the underlying layer. When the cone tip is in the lower layer, the development distance is defined as the greatest distance between the top of the underlying layer and the cone tip for which q^m in the underlying layer is still "affected" by the upper layer. Specific criteria for determining sensing and development distances are often not reported, but appear to represent some degree of visually apparent effects (e.g., perhaps a few percent). Sensing and development distances and the equations describing how q^m varies from q^t_{strong} to q^t_{weak} and from q^t_{weak} to q^t_{strong} are often assumed to be mirror images of each other (e.g., Xu and Lehane 2008, Mo et al. 2015), which is an approximation that implies equal influence of the soil resistance in front of and behind the cone tip. In reality, experimental and theoretical studies (e.g., Ahmadi and Robertson 2005, Tehrani et al. 2018) show that sensing distance in a strong layer over a weak layer is greater than the development distance in a strong layer under a weak layer, which indicates that soils in front of the cone tip have a greater influence on penetration resistance than the soils behind the cone tip. Sensing and development distances in a strong layer adjacent to a weak layer increase as the ratio of the soil layer strengths increases (e.g., Xu and Lehane 2008). Sensing and development distances in a weak soil adjacent to a stronger soil are smaller than those for the stronger soil, and they decrease as the ratio of the soil layer strengths increases. For example, Walker and Yu (2010) show that q^t for a soft clay layer embedded in stronger clay can be almost fully developed in layers as thin as 2–3 cone diameters.

Application of thin-layer or transition zone corrections in practice is relatively uncommon due to a number of challenges. Their application to field situations is generally subjective, such that the corrections applied can vary significantly between different individuals analyzing the same data. The ability to consistently distinguish between sharp interfaces and graded interfaces (e.g., upward or downward fining sequences) is uncertain. The procedures are difficult to automate, and time consuming to apply in the absence of an automated processing method. These challenges, combined with the fact that the results may only have a modest effect for many design/evaluation problems, appears to be why their use is relatively uncommon.

3 INVERSE FILTERING PROCEDURE

Inverse filtering is widely used in image and signal processing for a wide range of measurement applications (e.g., Cristobal et al. 2011). Inverse filtering can help restore or improve the quality of an image or measurement if a good model can be developed for the function that "blurred" the measurement and if the signal-to-noise ratio in the measurement is favorable. Inverse filtering often responds poorly to any noise present in the measurement because noise tends to be at high spatial frequencies and the blurred measurement tends to be weakest (i.e., have the lowest signal strength) at high spatial frequencies.

The application of inverse filtering techniques to cone penetration data begins with the assumption that there is a "true" cone penetration resistance that would be obtained if the cone penetration test data were dependent solely on the soil properties at a point (i.e., measured by an infinitely small cone without particle size effects). The measured cone penetration resistance depends on soil properties within a zone of influence around the cone tip, such that cone penetration acts as a low-pass spatial filter in sampling the true profile; i.e., strong variations in soil properties over short distances correspond to short wavelengths (or high spatial frequencies) that are masked or filtered by the cone penetration process. An inverted cone penetration resistance is obtained by applying an inverse cone penetration filter to the measured cone penetration resistance profile.

It is convenient to simplify notation for cone penetration resistances for the present purposes as follows. True, measured, and inverted cone penetration resistances are identified with superscripts as q^t, q^m, and q^{inv}, respectively. The q^m refers to the resistance after any correction for unequal area effects (i.e., pore pressure behind the cone tip) has already been applied, and all three terms are normalized by atmospheric pressure. The conventional subscripts for identifying cone penetration resistances as corrected for unequal area effects or normalized by atmospheric pressure (e.g., q_{tN}) are omitted for clarity.

The filtering effect of the cone penetration process can now be expressed as,

$$q^m(z) = q^t(z) * w_c(z) \quad (1)$$

where the asterisk indicates convolution of q^t with the cone penetration filter (w_c). Convolution refers to the integral of the point-wise multiplication of the two functions as a function of the amount that one of the functions is shifted relative to the other. The spatial filtering effect can alternatively be expressed as,

$$q^m(z) = \int_{z_{min}}^{z_{max}} q^t(\tau) w_c(z-\tau) d\tau \quad (2)$$

The convolution integration limits, which are $-\infty$ to ∞ for the general case, are set equal to the depth limits for the CPT sounding (i.e., z_{min} to z_{max}), which also ensures that the q^m vector retains the same length as the q^t vector.

The inversion of this filtering process is complicated by the strongly nonlinear nature of w_c and limitations on the highest spatial frequencies (shortest physical wavelengths) for which the inversion is meaningful. For cone penetration, the highest spatial frequencies for which the measurements may contain meaningful information could be governed by either the data sampling interval or the physical size of the cone. Higher spatial frequencies than are measureable can exist in the field, such as those associated with discrete jumps in q^t across interfaces, but these higher spatial frequencies almost certainly cannot be reliably inverted from the measurements. For this reason, the inverse filtering procedure must include steps that remove any of these higher spatial frequency components from the solution, as discussed later.

The inverse filtering procedure proposed herein has three primary components: (1) a model for how the cone penetrometer acts as a low-pass spatial filter in sampling the true distribution of soil properties versus depth, (2) a solution procedure for iteratively determining an estimate of the true cone penetration resistance profile from the measured profile and cone penetration filter model, and (3) a procedure for identifying sharp transition interfaces and correcting the data at those interfaces. The following sections address each of these three components.

4 CONE PENETRATION FILTER MODEL

Any cone penetration filter model needs to account for the primary influencing factors, recognizing that a perfect filter model with the full complexity of factors is not yet realizable. The proposed filter model for the current study is expressed as a function of the q^t profile alone, though additional information from f_s, u_{bt}, V_s, or other measurements may prove beneficial. The function forms and parameters for the current model were developed to be consistent with the body of results from prior studies regarding thin layer effects and sensing/development distances (Table 1), including their dependencies on various soil profile characteristics as discussed in the previous section.

The cone penetration filter (w_c) model, shown in Figure 3, is the normalized product of two functions, w_1 and w_2, as,

$$w_c = \frac{w_1 w_2}{\sum w_1 w_2} \quad (3)$$

where w_c, w_1, and w_2 are all functions of z', which is the depth relative to the cone tip normalized by the cone diameter (d_c),

$$z' = \frac{z - z_{tip}}{d_c} \quad (4)$$

where z_{tip} is the current depth of the cone tip. The dependence of w_c, w_1, and w_2 on z' is omitted from notation for simplicity.

The w_1 term accounts for the relative influence of any soil decreasing with increasing distance from the cone tip as,

$$w_1 = \frac{C_1}{1 + \left(\dfrac{z'}{z'_{50}}\right)^{m_z}} \quad (5)$$

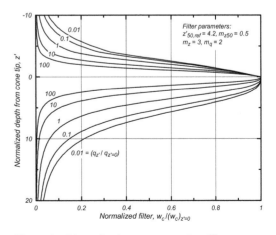

Figure 3. Normalized cone penetration filter versus normalized depth from the cone tip with lines for $q^t_{z'}/q^t_{z'=0} = 0.01, 0.1, 1, 10,$ and 100.

where z'_{50} is the normalized depth at which $w_1 = 0.5C_1$, and the exponent m_z is a parameter that adjusts the variation of w_1 with z'. The parameter C_1 is equal to unity for points below the cone tip, and linearly reduces to a value of 0.5 for points located more than 4 cone diameters above the cone tip as,

$$C_1 = 1 \quad for \quad z' \geq 0$$
$$= 1 + \frac{z'}{8} \quad for \quad -4 \leq z' < 0 \quad (6)$$
$$= 0.5 \quad for \quad z' < -4$$

The value of z'_{50} is computed as,

$$z'_{50} = 1 + 2(C_2 z'_{50,ref} - 1)\left(1 - \frac{1}{1 + \left(\frac{q^t_{z'=0}}{q^t_{z'}}\right)^{m_{50}}}\right) \quad (7)$$

where the product $C_2 z'_{50,ref}$ corresponds to the value of z'_{50} whenever $q^t_{z'}$ is equal to q^t at the cone tip (i.e., $q^t_{z'=0}$). The parameter C_2 is equal to unity for points below the cone tip, and less than unity (0.8 is used herein) for points above the cone tip. Thus, $z'_{50,ref}$ is the value of z'_{50} for points below the cone tip whenever $q^t_{z'}$ is equal to q^t at the cone tip.

The w_2 term adjusts the relative influence that soils away from the cone tip will have on the penetration resistance based on whether those soils are stronger or weaker than the soil immediately at the cone tip. If the soil at a given depth is weaker than the soil at the cone tip, its relative influence on the penetration resistance is increased, and vice versa if it is stronger than the soil at the cone tip. The w_2 term is computed as,

$$w_2 = \sqrt{\frac{2}{1 + \left(\frac{q^t_{z'}}{q^t_{z'=0}}\right)^{m_q}}} \quad (8)$$

where the exponent m_q is a parameter that adjusts the variation of w_2 with $q^t_{z'}/q^t_{z'=0}$.

The resulting filter model is shown in Figure 3 as w_c (normalized by w_c at the cone tip) versus z' for $q^t_{z'}/q^t_{z'=0}$ values of 0.01, 0.1, 1, 10, and 100 for the baseline set of parameters: $z'_{50,ref} = 4.0$, $m_z = 3.0$, $m_{50} = 0.5$, and $m_q = 2$. The soils above the cone tip receive about one half the weight received by the soils below the cone tip, all else being equal (i.e., comparing the relative areas under the w_c curves). For this parameter set, the filter model predicts that the measured penetration resistance will be controlled by soils within 2–3 diameters of the cone tip if they are far weaker than the soils further away (e.g., $q^t_{z'}/q^t_{z'=0} = 100$ at larger distances). Conversely, the measured penetration resistance will be significantly affected by soils as far as 15–20 diameters below the cone tip if the soils at these larger distances are far weaker than the soils near the cone tip (e.g., $q^t_{z'}/q^t_{z'=0} = 0.01$ at the larger distances). For more uniform soil profiles with only modest variations in q^t, the measured penetration resistance is controlled by soils within about 9 cone diameters of the tip. In addition, the asymmetry of the filter model is required to simulate the observed asymmetry in the sensing and development distances for strong layers embedded in weaker soil.

The filter convolution process is illustrated for a two-layer profile in Figure 4. For this illustration, the cone tip has reached the middle of a stronger layer embedded in a weaker soil profile, as depicted by the profile of q^t. The variation of w_c with depth is shown for that present cone tip depth. The q^t vector is point-wise multiplied by the w_c vector and the product summed to obtain the value of q^m for the present cone tip depth, which is shown as the single solid symbol on the right hand plot. This process is repeated for all other cone tip depths (with w_c shifting accordingly) to arrive at the profile of q^m shown as a dashed line on the right hand plot. For the cone tip at the middle of the stronger layer, the filter w_c decreases with distance from the cone tip in either direction for points in the stronger layer (since $q^t_{z'}/q^t_{z'=0} = 1$ throughout the layer), but steps up to larger values at the interface with the weaker soil because $q^t_{z'}/q^t_{z'=0} \ll 1$ for points in the weaker soil (which increases w_c as shown in Figure 3).

The numerical evaluation of the convolution integral involved two other details. First, the filter window was truncated at a length of 60 cone diameters, centered at the cone tip (i.e., $-30 \leq z' \leq 30$), for the examples presented in this paper. This truncation reduces computations and has minimal effect on results because w_c is close to zero at these distances from the cone tip. Second, the computation of w_c by Equation 3 is restricted to points that fall within the depth limits of the CPT sounding (i.e., the filter window is truncated at the data boundaries). This restriction ensures that the total area under the filter remains equal to unity near the upper and lower limits of the CPT sounding.

The behavior of the filter model is illustrated in Figure 5 showing q^t and q^m versus depth for three idealized soil profiles that each have four layers of a stronger soil interbedded in a weaker soil. The stronger soil has $q^t = 100$ in all cases, whereas the weaker soil has $q^t = 1$, 10, and 50 in Figures 5a, 5b, and 5c, respectively. The four stronger layers have thicknesses of 5, 10, 20, and 60 cone diameters in all cases. The digital data for these idealized profiles was generated with a uniform 20 mm sampling interval. The peak q^m that is computed

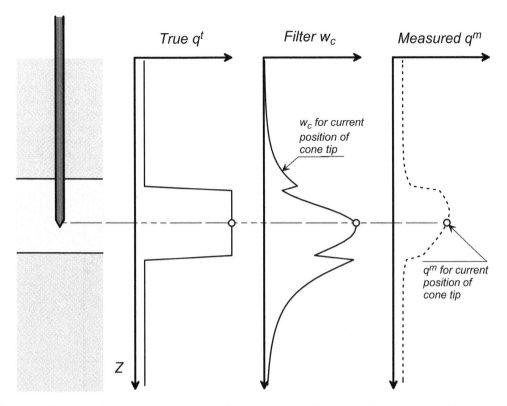

Figure 4. Illustration of the convolution of q^t with the cone penetration filter to obtain q^m at a given point in a layered profile.

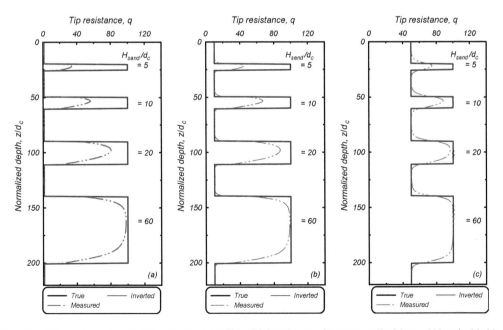

Figure 5. Values of q^t, q^m and q^{inv} for idealized profiles with interlayers of a strong soil with $q^t = 100$ embedded in a weaker soil with: (a) $q^t = 1$, (b) $q^t = 10$, and (c) $q^t = 50$. Note that q^{inv} almost perfectly overlays q^t.

to develop in the stronger layers decreases with decreasing layer thickness and decreasing strength in the surrounding soils.

Thin layer correction factors were subsequently derived using the above type of analysis for a uniform strong layer embedded in a uniform weaker deposit (e.g., Figure 1), with the results summarized in Figure 6. The derived values of K_H increase as q^t_{strong}/q^t_{weak} increases, but the rate of increase in K_H diminishes as q^t_{strong}/q^t_{weak} becomes larger. The K_H values approach unity for layers thicknesses greater than about 40 cone diameters, and are between 1.5 and 1.9 for layers about 10 cone diameters thick with q^t_{strong}/q^t_{weak} of 10–100. These K_H values are reasonably consistent with those of prior experimental and theoretical studies. The values of K_H increase rapidly as the thickness of the strong layer becomes less than about 5–10 cone diameters.

Sensing and development distances were similarly derived for an idealized two-layer system having q^t_{upper}/q^t_{lower} ratios ranging from 0.02 to 50. The sensing distance for the cone when it is in the upper layer (Δz_{sens}) and the development distance for the cone when it has entered the lower layer (Δz_{dev}), were determined as the distances from the interface where the difference between q^m and q^t was 5, 10, or 20%. The resulting values for $\Delta z_{sens}/d_c$ and $\Delta z_{dev}/d_c$ are plotted versus q^t_{upper}/q^t_{lower} in Figures 7a

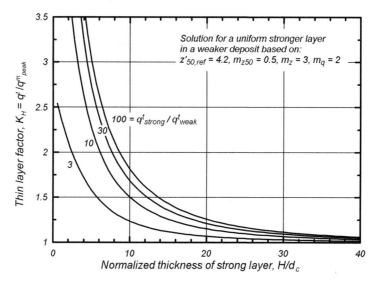

Figure 6. Thin layer correction factors computed for a uniform stronger layer of thickness H in a uniform weaker deposit.

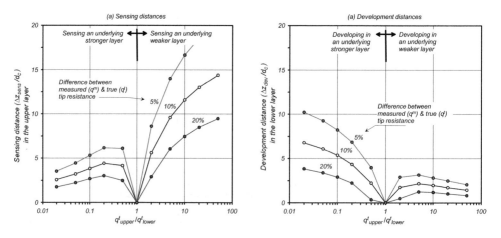

Figure 7. Sensing and development distances for a two layer system: (a) sensing distances, and (b) development distances.

and 7b, respectively. The $\Delta z_{sens}/d_c$ ranges from 2 to 6 when the cone is sensing an underlying stronger layer (left side of Figure 7a) and can be as large as 10–20 when the cone is sensing an underlying weaker layer (right side of Figure 7a). The $\Delta z_{dev}/d_c$ ranges from 2 to 10 when the cone is developing in an underlying stronger layer (left side of Figure 7b) and can be as small as 1 to 3 when the cone is developing in an underlying weaker layer (right side of Figure 7b).

The functional forms of the filter and the calibration of the filter parameters, as presented in this section, were iteratively developed to be consistent with the available experimental and theoretical data on thin layer and transition zone behaviors discussed in the previous section. The filter parameters can be adjusted to increase or decrease the thin layer effects or the sensing and development distances; for example, increasing z_{50ref} will cause the sensing/development distances to increase and the thin layer factors to increase for a given sand layer thickness. The parameter values adopted herein are considered reasonable values for general application, pending further experiences and validation studies.

5 SOLUTION PROCEDURE

5.1 Inversion for tip resistance

Inversion of q^m to obtain an estimate of q^t requires an iterative solution procedure because the filter is nonlinearly dependent on the unknown q^t. A method of successive substitutions as outline below was found to work well. The basic equations are first rearranged to obtain the difference between q^t and q^m as,

$$dq = q^t - q^m \quad (9)$$

where q^m is equal to the convolution of q^t with the filter, leading to

$$dq = q^t - q^t * w_c \quad (10)$$

The value of q^t can then be determined as,

$$q^t = q^m + dq \quad (11)$$
$$q^t = q^m + (q^t - q^t * w_c) \quad (12)$$

The above equation can now be solved by successive iterations as,

$$q_{n+1}^{inv} = q^m + (q_n^{inv} - q_n^{inv} * w_c) \quad (13)$$

where q_n^{inv} is the result of the nth iteration. The iteration process is initiated with the first estimate of q^{inv} set equal to q^m. Iterations continue until the following error criterion is satisfied,

$$err = \frac{\sum \left|(q_{n+1}^{inv} - q_n^{inv})_i\right|}{\sum \left|(q^m)_i\right|} < 10^{-6} \quad (14)$$

This solution procedure, without any adjustments, is not well constrained at spatial frequencies that are higher than justifiable based on the data sampling interval or the physical size of the cone. These higher-than-justifiable spatial frequency components can also impede convergence of the solution, as shown later. Two additional steps were therefore added to the solution procedure to remove these higher spatial frequencies and improve convergence; these additional steps are introduced in sequence below, after first illustrating performance of the solution procedure without either additional step.

Performance of the above solution procedure (without the additional steps) is illustrated in Figure 5, which shows an idealized q^t profile along with a q^m computed by convolving q^t with the baseline filter model. Also shown in these plots are the q^{inv} computed iteratively from q^m with the same or "true" filter. For the cases in Figures 5a and 5b, the solution procedure converged and the differences between q^{inv} and q^t are negligible. These cases illustrate how the solution procedure is essentially perfect if the filter is known perfectly and the solution converges. For the case in Figure 5c, the solution procedure did not converge, but rather stabilized at a solution where q^{inv} included a small amount of low-level, high spatial frequency (short wavelength) noise in the stronger layers (barely visible as ripples in the q^{inv}). The wavelength of this noise is similar to the cone tip length, which is when the solution process can become less stable.

The potential effects of "noise" or the potential challenges with inverting information from very thin interlayers using the above solution procedure is further illustrated in Figure 8(a). This figure shows a profile of uniform soil with $q^m = 10$ except for thin intervals that are 1.1, 1.7, 2.2, and 3.4 cone diameters thick and have $q^m = 12$. The digital data were generated with a uniform 20 mm sampling interval, such that these depth intervals correspond to 2, 3, 4, and 6 data points with $q^m = 12$. Inversion of this q^m profile did not converge (i.e., meet the error criteria), but did stabilize at the solution shown in Figure 8a with peak q^{inv} values in the range of 30–40. These high q^{inv} values are sensitive to details of the filter model and are not reliable.

The solution procedure therefore requires an additional smoothing step that removes some of the highest spatial frequencies (shortest wavelengths) during the inversion process. After each

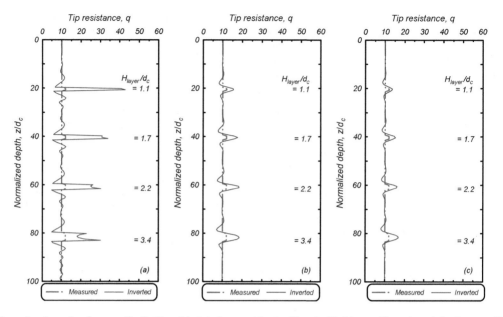

Figure 8. Inversion for a profile that has thin interlayers with $q^m = 12$ embedded in a uniform deposit having $q^m = 10$: (a) results using the first solution procedure, (b) results for the solution procedure with smoothing over the cone tip length, and (c) results for the solution procedure with smoothing followed by low-pass spatial filtering.

iteration, q^{inv} is smoothed (or low-pass spatial filtered) by taking a moving average over a window of a specified number of data points. The smoothing window is defined as the larger of either three points or the ceiling (rounding up) of the cone tip length (i.e., 0.866 d_c) divided by the data point depth spacing (i.e., Δz). The first criterion governs for a standard 10 cm² cone and uniform 20 mm data sampling interval, whereas the second criterion governs if the data sampling interval is much smaller (e.g., 5 mm). In this regard, the solution procedures were tested for data sampling intervals of 1–50 mm to ensure they continued to perform well for non-standard conditions. The inclusion of this smoothing step removes spatial frequencies that cannot be reliably inverted, greatly improves the convergence rate, and eliminated any cases of non-convergence. For example, application of the revised solution procedure to the profile shown in Figure 8 produced the smoother and more reasonable results shown in Figure 8b. In addition, the error term is plotted versus iteration number in Figure 9 for the solution procedures used for Figures 8a and 8b, illustrating how the inclusion of smoothing greatly improved the numerical performance of the solution procedure.

A second low-pass spatial filtering step was added to the solution procedure, although its effects are small for most situations. For this last step, the converged q^{inv} from the inversion with smoothing is convolved with another low-pass filter,

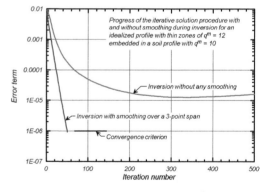

Figure 9. Progress of the iterative soil procedure with and without smoothing during inversion.

$$q^{inv} = q^{inv} * w_{c2} \qquad (15)$$

where w_{c2} is the same filter model as w_c except that z_{50ref} is reduced to the length of the cone tip (i.e., 0.866 d_c). The application of this additional step in the inverse filtering of the profile shown in Figure 8 produced the slightly smoother result shown in Figure 8c. The results in Figure 8c illustrate that the overall solution procedure with smoothing during inversion followed by low-pass spatial filtering appears well-suited for handling high spatial frequency noise.

Performance of the overall solution procedure (inversion with smoothing followed by low-pass spatial filtering) is illustrated in Figure 10 for the same idealized profiles examined previously (in Figure 5). The q^{inv} profiles still reasonably approximate the q^t profiles, though slightly rounded at the interfaces and slightly smaller than q^t for the interlayers that are only 5 cone diameters thick.

Performance of the overall solution procedure can also be expressed in terms of the "net" thin-layer factor that it produces for idealized two-layer systems, as shown in Figure 11. These net factors are computed as the peak q^{inv} in a thin layer divided by the peak q^m for that same layer, and thus are

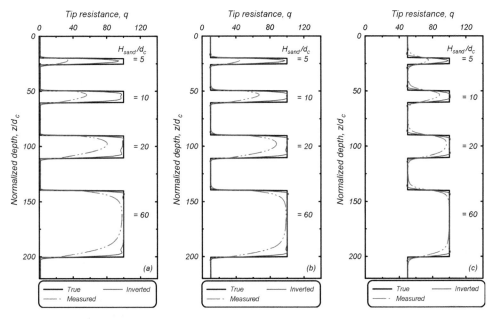

Figure 10. Values of q^t, q^m and q^{inv} for three idealized profiles using the solution procedure with smoothing followed by low-pass spatial filtering. Stronger layers have $q^t = 100$ and the weaker layers have (a) $q^t = 1$, (b) $q^t = 10$, and (c) $q^t = 50$.

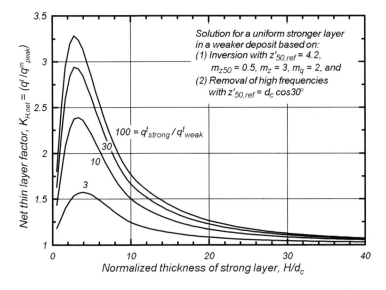

Figure 11. Net thin layer correction factors computed using inversion with low-pass spatial filtering.

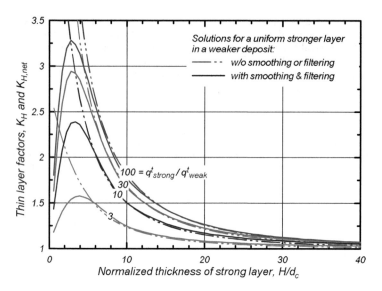

Figure 12. Comparison of the thin layer correction factors from the solutions with and without the smoothing and low-pass spatial filtering steps.

illustrative of how much amplification the inversion process may apply to the q^m in sand layers of various thicknesses. The $K_{H,net}$ curves for a range of q^t_{strong}/q^t_{weak} ratios all reach their peak values at strong-layer thicknesses of 3–4 cone diameters, and decrease toward unity with decreasing layer thickness. The fact $K_{H,net}$ tends to unity as the strong-layer thickness tends to zero reflects the fact that cone penetration measurements cannot provide meaningful information on soil properties when they vary strongly across interlayers that are less than about 2 or 3 cone diameters thick. The previously presented K_H values are overlain with the $K_{H,net}$ values in Figure 12, illustrating how the K_H values rapidly increase toward large values at these small layer thicknesses, which relates to why inversion without smoothing or low pass spatial filtering performed poorly. The solution procedure with smoothing during inversion followed by low-pass spatial filtering produces equivalent $K_{H,net}$ values that are reasonable for strong-layer thicknesses that are greater than about 3–4 cone diameters and appropriately conservative for thinner layers where inversion is not reliable.

5.2 Inversion for sleeve friction

The inversion of sleeve friction (f_s) profiles could conceptually follow a similar process as used for tip resistance. The filter model for f_s could have a different form and distribution. For example, examination of field data illustrates that f_s is more sensitive to thin layers than q_t, which would imply a smaller zone of influence or greater weighting of data near the sleeve. Unfortunately, there are currently limited data available to guide development of a separate filter model for f_s.

The strategy adopted herein was to develop f_s^{inv} values from the q^{inv} values. This procedure assumes that the pairs of normalized tip resistance (Q) and normalized sleeve friction ratio (F) for the inverted and measured data lie along a radial line originating from the origin of the Soil Behavior Type Index (I_c). The values of Q and F are computed as,

$$Q = \left(\frac{q-\sigma_{vc}}{P_a}\right)\left(\frac{P_a}{\sigma'_{vc}}\right)^n \quad (16)$$

$$F = \left(\frac{f_s}{q-\sigma_{vc}}\right)100\% \quad (17)$$

where σ_{vc} is the total vertical stress, σ'_{vc} is the effective vertical stress, and P_a is atmospheric pressure. The stress exponent n is computed following the relationship recommended by Robertson (2009),

$$n = (0.381I_c + 0.05\sigma'_{vc} - 0.15) \leq 1 \quad (18)$$

and the value of I_c is computed using the form recommended by Robertson and Wride (1997),

$$I_c = \left[(3.47 - \log(Q))^2 + (1.22 + \log(F))^2\right]^{0.5} \quad (19)$$

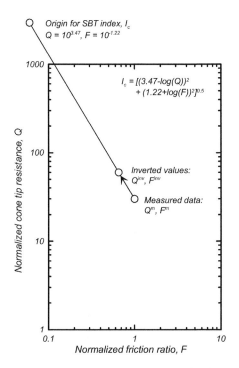

Figure 13. Procedure used to compute F^{inv} given Q^m and F^m for the field measurements and the value of Q^{inv} obtained from the inversion procedure.

The I_c value can be visualized as the radial distance between any Q-F pair and an origin point located at $Q = 2,951$ and $F = 0.0603$, as shown in Figure 13. The measured q^m and f_s^m values are used to compute Q^m and F^m values, and the inverted q^{inv} is used to compute a Q^{inv} value. The value of F^{inv} is then computed based on the assumption that the inverted point moved radially with respect to the I_c origin,

$$F^{inv} = 10^{\left(\frac{(3.47-\log(Q^{inv}))}{(3.47-\log(Q^m))}(1.22+\log(F^m))-1.22\right)} \quad (20)$$

The above procedure is schematically illustrated for a single data point in Figure 13. The value of f_s^{inv} is then computed from F^{inv} using the definition of F.

6 INTERFACE DETECTION AND CORRECTION

The detection of sharp interfaces and correction of the data at those interfaces is a separate step because the inversion process does not reproduce the high-spatial frequencies associated with a step in the q^t profile (i.e., the smoothing during inversion followed by low-pass spatial filtering removes those spatial frequencies because they cannot be inverted reliably). Identifying sharp transitions is complicated by the fact that some interfaces may be sharp and others may be graded (e.g., upward fining sequences), and thus the criteria and process will involve some subjectivity and should be subject to confirmation by borehole sampling data or knowledge of local geology.

A sharp transition (or interface) is considered likely to exist if the rate of change in the logarithm of q^{inv} with respect to normalized depth is larger than a specified criterion. The rate of change across each discrete data sampling interval is computed as,

$$m_i = \frac{\ln(q_{i+1}^{inv}) - \ln(q_i^{inv})}{z'_{i+1} - z'_i} \quad (21)$$

Note that the index for m corresponds to the intervals between measurement points; i.e., if there are N points in the q^{inv} vector, there are N-1 points in the m vector. The value of m that would be consistent with the existence of a sharp interface can be estimated using experimental and theoretical estimates of the sensing and development distances. For example, if q changes by a factor of more than 2 over a distance of about 6 cone diameters (214 mm for a standard 10 cm² cone), then m will exceed 0.12 on average. On the other hand, if q smoothly changes by a factor of less than 10 over a graded interval that is about 1 m thick (or about 28 standard cone diameters), then m may not exceed 0.08 on average. Thus, a sharp interface may be expected to exist if m exceeds a specified threshold, m_t. A reasonable value for m_t, as adopted herein, may be on the order of 0.1, subject to refinement based on site-specific sampling and geologic information.

The transition zone over which q^{inv} at a sharp interface is still blurred will also have smoothly varying m values. The m values both above and below the point where the maximum m is obtained will be smaller than the maximum m for that transition zone. For the present study, the transition zone is assumed to include any contiguous measurement points where m is greater than $m_t/5$. If the zone identified by the above criteria is less than 3 cone diameters thick, then is it not considered a transition zone. If the zone identified by the above criteria is greater than 12 cone diameters thick where q^{inv} is increasing, then the interval is truncated to 12 cone diameters thick (centered on the original zone). If the zone identified by the above criteria is greater than 18 cone diameters thick where q^{inv} is decreasing, then the interval is truncated to

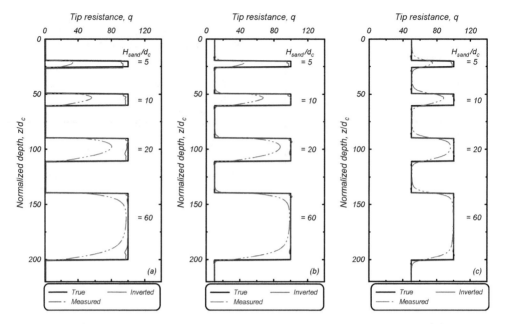

Figure 14. Values of q^t, q^m and q^{inv} for three idealized profiles using inversion with smoothing followed by low-pass spatial filtering and interface detection and correction. Stronger layers have $q^t = 100$ and the weaker layers have (a) $q^t = 1$, (b) $q^t = 10$, and (c) $q^t = 50$.

18 cone diameters thick (centered on the original zone). Once the transition zone has been defined, the q^{inv} and f_s^{inv} values at the top of the transition zone are assigned to all points in the upper 40% of the transition zone for penetration advancing into a stronger layer or the upper 60% of the transition zone for penetration advancing into a weaker layer. The q^{inv} and f_s^{inv} values at the bottom of the transition zone are assigned to all remaining points in the lower part of the transition zone.

Application of the interface detection and correction algorithm is illustrated in Figure 14 for the same idealized profiles presented in Figures 5 and 10. The interface detection and correction algorithm does not perfectly remove the transition zone, since it does not capture the slow rates of change in q at the start and end of each transition zone for this idealized profile. A greater portion of the interface could be removed by reducing m_t, but that can result in falsely identifying transition zones or computing overly thick transition zones when processing real CPT soundings.

7 INVERSION EXAMPLES FOR FIELD DATA

7.1 Applications using the baseline model

Application of the proposed inverse filtering procedure is illustrated in Figures 15–18 for cone soundings from four different sites. The inversion, low-pass spatial filtering, and interface detection steps are applied using the same baseline parameters specified in the previous sections. The first example (Figure 15) is CPT UC-4 along Sandholdt Road in Moss Landing, CA, USA where liquefaction-induced ground deformations were observed after the 1989 $M_w = 6.9$ Loma Prieta earthquake (Boulanger et al. 1997). The second example (Figure 16) is CPT 1–24 along Çark Canal in Adapazari, Turkey where no surficial evidence of liquefaction was observed after the 1999 $M_w = 7.5$ Kocaeli earthquake (Youd et al. 2009). The third example (Figure 17) is CPT 45185 at St. Theresa's school in Riccarton, Christchurch, New Zealand where no surficial evidence of liquefaction was observed during the 2010–2011 Canterbury Earthquake Sequence (Cox et al. 2017). The fourth example (Figure 18) is a CPT at Hinode Minami Elementary School, Urayasu, Japan where no evidence of liquefaction was observed following the 2011 $M_w = 9.0$ Tohoku earthquake (Cox et al. 2013). The data sampling intervals for these four CPT soundings were 50, 20, 10, and 25 mm, respectively. The latter three sites are of interest because some degree of liquefaction would have been expected based on common engineering liquefaction evaluation procedures, as reported in the various references. The details of liquefaction evaluations for each site are beyond the scope of the present paper, except to note that thin-layer and transition effects are considered to be one

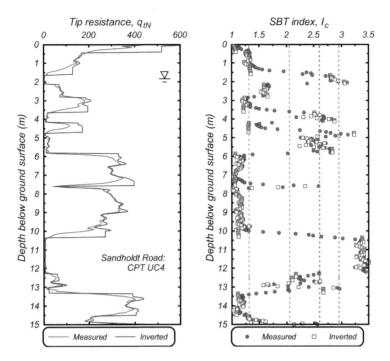

Figure 15. Measured and inverted q_{tN} and I_c profiles for CPT UC-4 at Sandholdt Road.

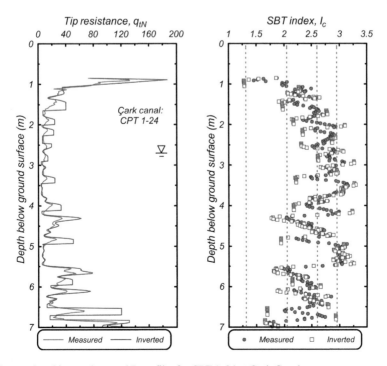

Figure 16. Measured and inverted q_{tN} and I_c profiles for CPT 1–24 at Çark Canal.

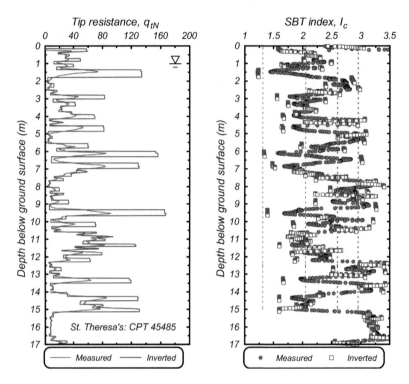

Figure 17. Measured and inverted q_{tN} and I_c profiles for CPT-45185 at St. Theresa's School.

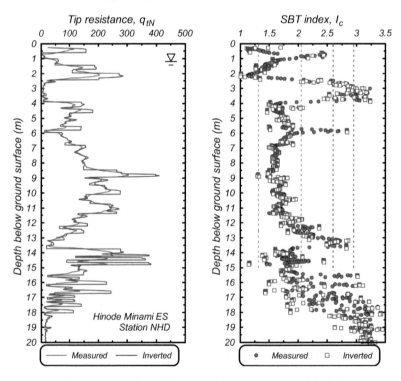

Figure 18. Measured and inverted q_{tN} and I_c profiles for Station NHD at Hinode Minami Elementary School.

of several factors that can contribute to the overprediction of liquefaction effects (e.g., Boulanger et al. 2016).

Examination of these four examples illustrates that the proposed CPT inversion procedure is performing as expected. The inversion process has negligible effects on q or I_c in intervals where the soil type or penetration resistance is relatively uniform. The inversion procedure has the greatest effect where well-defined thin-layers of stronger material are detected. The inversion procedure increases q^{inv} (relative to q^m) by factors up to 2–3 even for the thinnest strong interlayers at the highly interbedded sites, which reflects the upper limits that the present inversion parameters will produce (e.g., as illustrated by the net thin layer factors in Figure 11) and that can be justified given conventional cone diameters. The inversion procedure was

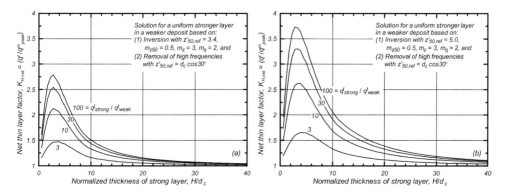

Figure 19. Net thin layer correction factors using the baseline set of filter parameters except for: (a) $z'_{50,ref} = 3.4$, and (b) $z'_{50,ref} = 5.0$

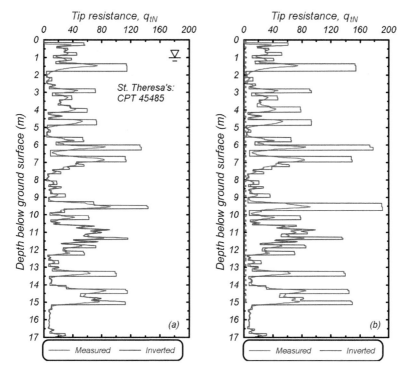

Figure 20. Measured and inverted q_{tN} profiles for CPT-45185 at St. Theresa's School using the baseline set of filter parameters except for: (a) $z'_{50,ref} = 3.4$, and (b) $z'_{50,ref} = 5.0$.

effective in removing transition zones at clearly identifiable contacts, while the detection criteria left other zones of slowly increasing penetration resistance unmodified.

7.2 Example using alternative filter parameters

The sensitivity of the inverse filtering procedure to alternative sets of filter parameters was also evaluated, recognizing that there is uncertainty in the filter model and individual parameters. The effect of individual parameter variations were examined in terms of how they affected the solutions for idealized problems (e.g., Figure 14), sensing and development distances for idealized two layer profiles (e.g., Figure 7), net thin layer correction factors (e.g., Figure 11), and results for field CPT soundings. Alternative sets of parameters that are consistent with the body of available experimental and theoretical data can be developed, including sets that provide more or less aggressive corrections to measured data.

For example, consider the effect of setting $z'_{50,ref}$ equal to 3.4 or 5.0 ($\pm$ 19% from the baseline value of 4.2), while keeping all other parameters equal to the baseline values. The corresponding $K_{H,net}$ values, as shown in Figures 19a and 19b, are smaller for $z'_{50,ref}$ = 3.4 and larger for $z'_{50,ref}$ = 5.0, with the differences being larger for smaller layer thicknesses. For a strong layer thickness of 3–5 cone diameters, using $z'_{50,ref}$ = 3.4 reduced the $K_{H,net}$ values by 8–18% (relative to those for the baseline set of filter parameters) whereas using $z'_{50,ref}$ = 5.0 increased the $K_{H,net}$ values by 6–18%.

The effects of using the above $z'_{50,ref}$ values on the inverse filtering of a field CPT sounding are illustrated in Figures 20a and 20b for CPT 45185 at St. Theresa's school in Riccarton. The use of $z'^{*}_{50,ref}$ = 3.4 (Figure 20a) resulted in smaller q^t values in the numerous thin sand interlayers relative to those for the baseline case (Figure 17), whereas use of $z'_{50,ref}$ = 5.0 (Figure 20b) resulted in larger q^t values. The differences in the inversion results for these two cases are consistent with the differences in the corresponding net thin layer correction factors.

8 CONCLUDING REMARKS

An inverse filtering procedure for developing estimates of the "true" cone penetration tip resistance and sleeve friction values from measured cone penetration test data was presented. The inverse filtering procedure has three primary components: (1) a model for how the cone penetrometer acts as a low-pass spatial filter in sampling the true distribution of soil resistance versus depth, (2) a solution procedure to iteratively determine an estimate of the true cone penetration resistance profile from the measured profile given the cone penetration filter model, and (3) a procedure to identify sharp transition interfaces and correct the data at those interfaces. The inverse filtering procedure was shown to produce equivalent thin layer factors and sensing/development distances that are consistent with the results of prior experimental and theoretical studies. Example applications of the inverse filtering procedure to four CPT soundings illustrated that the model performs well for a range of soil profile characteristics. The inverse filtering procedure was shown to provide an objective, repeatable, and automatable means for correcting cone penetration test data for thin-layer and transition zone effects.

Further experience with application of the inverse filtering procedure can be expected to lead to improvements in components of the procedure or improved guidance for its use in practice. Issues worth further examination include how f_s and q_c may filter differently, how the filter model might depend on other specific properties (stiffness, strength, drainage conditions) or stratigraphic sequences, and how the inversion process or parameters might be adjusted based on site-specific borehole and geologic information. The present model was developed with a focus on liquefaction problems, whereas applications to other problems may identify different aspects that warrant improvement. Regardless of advances, there will be practical limits to how well inverse filtering procedures, or any other technique, can correct cone penetration test data given the nature of the cone penetration process, the infinite possible variations in geologic conditions, the presence of noise in measurements, and other complex processes that influence the measurements (e.g., partial drainage, physical dragging of soil along with the cone). In this regard, the proposed inverse filtering procedure is intended to improve or enhance the field measurements, while recognizing that any inversion process will be neither unique nor perfect.

ACKNOWLEDGMENTS

The authors appreciate the financial support of the National Science Foundation (award CMMI-1635398) and California Department of Water Resources (contract 4600009751) for different aspects of the work presented herein. Any opinions, findings, conclusions, or recommendations expressed herein are those of the authors and do not necessarily represent the views of these organizations. Dr. Mohammad Khosravi and Mr. Mathew Havey contributed to the literature review and evaluation of alternative procedures for evaluating thin-layer effects. Dr. Dan

Wilson provided valuable comments on numerical aspects of inverse filtering procedures. Professor Brady Cox provided valuable review comments that improved the clarity of the paper. The authors appreciate the above support and interactions.

REFERENCES

Ahmadi, M. M. and Robertson, P. K. 2005. Thin-layer effects on the CPT q_c measurement. Canadian Geotechnical Journal, 42(5), 1302–1317.

Boulanger, R. W., Mejia, L. H., and Idriss, I. M. 1997. Liquefaction at Moss Landing during Loma Prieta Earthquake. Journal of Geotechnical and Geoenvironmental Engineering, ASCE, 123(5), 453–467.

Boulanger, R. W., Moug, D. M., Munter, S. K., Price, A. B., and DeJong, J. T. 2016. Evaluating liquefaction and lateral spreading in interbedded sand, silt, and clay deposits using the cone penetrometer. Geotechnical and Geophysical Site Characterisation 5, B. M. Lehane, H. Acosta-Martinez, and R. Kelly, eds., Australian Geomechanics Society, Sydney, Australia, ISBN 978-0-9946261-2-7.

Canou, J. 1989. Piezocone et liquefaction des sables. Rapport de synthese des travaux realizes au CERMES, Re-search Report CERMES/ENPC, Paris, 176 p.

Cox, B. R., Boulanger, R. W., Tokimatsu, K., Wood, C., Abe, A., Ashford, S., Donahue, J., Ishihara, K., Kayen, R., Katsumata, K., Kishida, T., Kokusho, T., Mason, B., Moss, R., Stewart, J., Tohyama, K., and Zekkos, D. 2013. Liquefaction at strong motion stations and in Urayasu City during the 2011 Tohoku-Oki earthquake. Earthquake Spectra, EERI, 29(S1), S55-S80.

Cox, B. R., McLaughlin, K. A., van Ballegooy, S., Cubrinovski, M., Boulanger, R. W., and Wotherspoon, L. 2017. In-situ investigation of false-positive liquefaction sites in Christchurch, New Zealand: St. Teresa's School case history. Proc., Performance-based Design in Earthquake Geotechnical Engineering, PBD-III Vancouver, M. Taiebat et al., eds., ISSMGE Technical Committee TC203, paper 265.

Cristobal, G., Schelkens, P., and Thienpont, H. 2011. Optical and digital image processing: Fundamentals and applications. Wiley-VCH Verlag GmbH & Co KGaA, Editors: G. Cristobal, P. Schelkens, and H. Thienpont, DOI: 10.1002/9783527635245.

Foray, P. & Pautre, J-L. 1988. Piezocone et liquefaction des sables: synthese des essais sur sites en Nouvelle-Zelande et des essais en Chambre de Calibration a l'IMG, Research Report, IMG, Grenoble, 70 p.

Hird, C. C., Johnson, P., and Sills, G. C. 2003. Performance of miniature piezocones in thinly layered soils. Géotechnique, 53(10), 885–900.

Lunne, T., Robertson, P.K., and Powell, J.M. 1997. Cone penetration testing in geotechnical practice. Blackie Academic & Professional, London, UK.

Mayne, P. 2007. Cone Penetration Testing—A synthesis of highway practice. NCHRP Synthesis 268, Transportation Research Board, Washington, D.C.

Meyerhof, G. G. and Valsangkar, A. J. 1977. Bearing capacity of piles in layered soils. Proc. 9th Int. Conf. Soil Mech. Found. Engng, Japan 1, 645–650.

Mlynarek, Z., Gogolik, S., and Poltorak, J. 2012. The effect of varied stiffness of soil layers on interpretation of CPTU penetration characteristics. Archives of Civil and Mechanical Engineering, 12, 253–264.

Mo, P.-Q., Marshall, A. M., and Yu, H.-S. 2017. Interpretation of cone penetration test data in layered soils using cavity expansion analysis. J. Geotechnical and Geoenvironmental Eng., 143(1), 10.1061/(ASCE) GT.1943-5606.0001577.

Mo, P. Q., Marshall, A. M., and Yu, H. S. 2015. Centrifuge modelling of cone penetration tests in layered soils. Géotechnique, 65(6), 468–481.

Munter, S. K., Boulanger, R. W., Krage, C. P., and DeJong, J. T. 2017. Evaluation of liquefaction-induced lateral spreading procedures for interbedded deposits: Cark Canal in the 1999 M7.5 Kocaeli earthquake. Geotechnical Frontiers 2017, Seismic Performance and Liquefaction, Geotechnical Special Publication No. 281, T. L. Brandon and R. J. Valentine, eds., 254–266.

Robertson, P. K. 2009. Interpretation of cone penetration tests—a unified approach. Canadian Geotechnical Journal, 46, 1337–1355.

Robertson, P. K. and Fear, C. E. 1995. Liquefaction of sands and its evaluation. Proc., 1st Int. Conf. on Earthquake Geotechnical Engineering, Ishihara, K. (ed), A. A. Balkema.

Robertson, P. and Wride, C., 1998. Evaluating cyclic liquefaction potential using the cone penetration test. Canadian Geotechnical Journal, 35(3), 442-459.

Sayed, S. M., and Hamed, M. A. 1987. Expansion of cavities in layered elastic system. International Journal of Numerical and Analytical Methods in Geomechanics, 11(2), 203–213.

Silva, M. F., and Bolton, M. D. 2004. Centrifuge penetration tests in saturated layered sands. Proc. ISC-2 on Geotechnical and Geophysical Site Characterization, Viana da Fonseca and Mayne (eds), 377–384.

Tehrani, F. S., Arshad, M. I., Prezzi, M., and Salgado, R. 2018. Physical modeling of cone penetration in layered sand. Journal of Geotechnical and Geoenvironmental Engineering, 144(1), 04017101, DOI:10.1061/(ASCE)GT.1943-5606.0001809.

Treadwell, D.D. 1976. The influence of gravity, prestress, compressibility, and layering on soil resistance to static penetration. Ph.D. thesis, University of California at Berkeley, Berkeley, Calif.

Van den Berg, P., De Borst, R. & Huetink, H. 1996. An Eulerian finite element model for penetration in layered soil. International Journal of Numerical and Analytical Methods in Geomechanics, 20, 865–886.

Vreugdenhil, R., Davis, R. and Berrill, J. 1994. Interpretation of cone penetration results in multilayered soils. International Journal of Numerical and Analytical Methods in Geomechanics, 18(9), 585–599.

Walker, J. and Yu, H.-S. 2010. Analysis of the cone penetration test in clay. Geotechnique, 60(12), 939–948.

Xu, X. T., and Lehane, B. M. 2008. Pile and penetrometer end bearing resistance in two-layered soil profiles. Géotechnique, 58(3), 187–197.

Youd, T. L., Idriss, I. M., Andrus, R. D., Arango, I., Castro, G., Christian, J. T., Dobry, R., Finn, W. D. L., Harder, L. F., Hynes, M. E., Ishihara, K., Koester, J. P., Liao, S. S. C., Marcuson, W. F., Martin, G. R., Mitchell, J. K., Moriwaki, Y., Power, M. S., Robertson, P. K., Seed, R. B., and Stokoe, K. H. 2001. Liquefaction resistance of soils: summary report from the

1996 NCEER and 1998 NCEER/NSF workshops on evaluation of liquefaction resistance of soils, J. Geotechnical and Geoenvironmental Eng., ASCE 127(10), 817–33.

Youd, T. L., DeDen, D. W., Bray, J. D., Sancio, R., Cetin, K. O. and Gerber, T. M. 2009. Zero-displacement lateral spreads, 1999 Kocaeli, Turkey, earthquake. Journal of Geotechnical and Geoenvironmental engineering, 135(1), 46–61.

Yue, Z.Q., and Yin, J.H. 1999. Layered elastic model for analysis of cone penetration testing. International Journal for Numerical and Analytical Methods in Geomechanics, 23, 829–843.

Use of CPT for the design of shallow and deep foundations on sand

K.G. Gavin
Delft University of Technology, Delft, The Netherlands

ABSTRACT: A wide range of empirical correlations linking the Cone Penetration Test (CPT) end resistance q_c and the resistance of shallow and deep foundations in sand have been published. Both National and European organisations are attempting to introduce standard methods into practice that unify these approaches. In this paper experimental data and finite element analyses are reviewed to examine the mechanisms governing foundation behaviour in a bid to move towards those unified approaches. For shallow foundations and non-displacement piles, sand creep was found to affect correlations between q_c and the mobilised bearing resistance. For pile foundations direct correlations between q_c and pile end resistance that were dependent only on pile installation method are reported. In the case of shaft resistance, constant correlation factors between q_c and average shaft resistance are possible for non-displacement piles. For the case of displacement piles, correlations that include the effects of friction fatigue are recommended.

1 INTRODUCTION

Whilst conventional bearing capacity and earth pressure approaches are widely used to design shallow and deep foundation in sand, many design codes are moving towards Cone Penetration Test (CPT) based design methods. This is as a result of a significant research effort in the area of foundation design in recent years. An issue facing both designers operating internationally and causing debate for those drafting unified codes such as Eurocode 7, is that many national recommendations have been published linking CPT end resistance, q_c with the bearing resistance of shallow foundations and piles. However, these values are rarely consistent, and in some cases can vary significantly. Whilst some of these differences may be caused by geology, pile types adopted, and a wide range of methods (averaging techniques) for estimating design q_c profiles, it arises at least in part due to a lack of understanding of the mechanisms controlling foundation behavior. Even internally within some countries, there is ongoing debate surrounding CPT based design.

Recent updates to the Dutch design code have caused significant debate within the geotechnical engineering community. Based on an interpretation of static load test data from across the Netherlands, CUR report 229 proposed a 30% reduction to the pile base reduction factor that will result in larger and longer piles from January 1st 2017. These changes are largely due to a lack of reliable pile load tests data in the Netherlands as the country is unusual in that it does not routinely test the axial capacity of piles. This has been met with some resistance from industry since historical pile failure rates are very low.

In this paper, the results of experiments performed on foundations in sand and finite element analyses are compiled in an attempt to provide an insight into factors that may influence correlations between CPT q_c and the behavior of both shallow and deep foundations in sand.

2 SHALLOW FOUNDATIONS

2.1 *General description*

Routine design of shallow footings on sand is often undertaken using the conventional bearing capacity approach to calculate the ultimate bearing resistance (q_{ult}):

$$q_{ult} = 0.5\gamma\ B\ S_\gamma\ d_\gamma\ N_\gamma + \gamma D\ S_q d_q N_q \qquad (1)$$

where: = unit weight of Soil, B = foundation width, D = foundation depth, $S_\gamma, S_q\ d_\gamma$ and d_q are factors to account for footing shape and depth, whilst N_γ and N_q are bearing capacity factors which depend on the friction angle (ϕ') of the soil and the footing shape.

Additional factors can be included to take account of situations such as sloping ground conditions and inclined or eccentric loading, although there is a strong argument that the latter should be treated using interaction diagrams as proposed by Gottardi and Butterfield (1993) and Houlsby and Cassidy (2002). Designers face key challenges in the application of Equation (1). These include the choice of appropriate shape and depth factors, and most importantly bearing capacity factors:

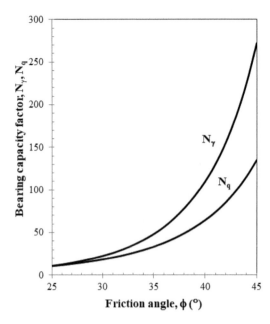

Figure 1. Typical bearing capacity factors for shallow footings.

Although Randolph et al. (2004), note that the accuracy with which bearing capacity factors can be determined has increased substantially in recent years, it is clear from Figure 1 that the values increase substantially over the range of soil strength typically encountered. Because of the difficulties in obtaining high quality samples, coupled with the challenges of interpreting soil strength parameters for the complex stress and strain paths experienced during loading at the relatively low stress levels applicable to footing design, greater reliance is placed on using empirical formulae linking ϕ' and q_c or using direct correlations linking q_c and the soil bearing resistance.

2.2 CPT based methods for shallow foundation capacity

In addition to the challenges outlined above, a further driver towards the use of in-situ test data in shallow foundation design was identified by Briaud (2007) from interpretation of field test data. Briaud and Gibbens (1999) report tests performed on square footings where the footing width, B varied from 1 m to 3 m. The footings were founded 0.75 m below the ground surface in recently deposited medium-dense sand. The site was in a lightly over-consolidated state (OCR ≈ 2) following the removal of about 1.0 m overburden. The mean CPT q_c resistance ranged from 5 to 7.25 MPa in the zone of influence of the footings. The pressure-settlement response (including un-load and re-load loops) of three of the footings is illustrated in Figure 2a. It is apparent that for a given settlement s, the mobilised bearing pressure, q reduced with increasing footing width. Another feature of the test was that for values of q greater than 650 kPa, significant creep settlement occurred during maintained load steps.

When the mobilised pressure was normalised by the q_c value averaged over the zone of influence, and he footing settlement divided by the footing width, a relatively unique normalised pressure-settlement response was obtained for the site, see Figure 2b. Noting that the unload-reload portions

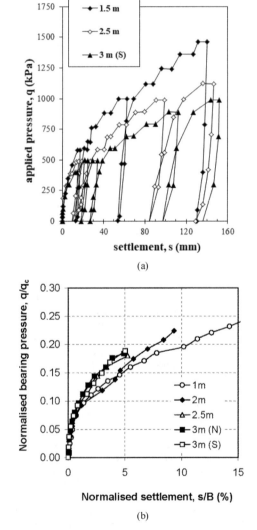

Figure 2. (a) Pressure-Settlement response for three footings (b) normalised pressure-settlement response.

of the data were removed for clarity and the maximum settlement for each load step were plotted, as it is clear from Figure 2a that the shape of the curve is affected by the load test procedure. Adopting a definition of failure as being the pressure mobilised when the footing settlement reached 10% of the footing width, $q_{0.1}$ the data suggests the adoption of a direct method for estimation of the footing resistance of the form given in Equation 2 with an α value of 0.20.

$$q_{0.1} = \alpha\, q_c \qquad (2)$$

Briaud (2007) considered data from other test sites to investigate the effect of footing depth, D and found that the normalised pressure-settlement response was independent of both footing width and relative embedment (D/B). He suggests that whilst Equation 1 would produce good estimates of q_{ult} of footings in soil profiles where the soil strength increased linearly with depth, in over-consolidated sand or in deposits where the near surface soil is unsaturated, the soils strength is often relatively constant with depth and the assumption that q_{ult} increases with footing width, B or footing depth, D is not valid.

Based on a review of footing tests from the literature, Eslaamizaad and Robertson (1996) found that the back-figured α values varied with soil density, relative embedment and footing shape. Randolph et al. (2004) summarised the results of laboratory tests, field tests and numerical analyses performed on shallow footings and buried piles. They report a relatively wide range of α values varying from 0.13–0.21. However, there was no evidence that α varied with footing width or sand state.

Finite Element (FE) analyses present an ideal environment in which to consider the mechanisms governing footing response and to perform sensitivity analyses for foundations. Lee and Salgado (2005) reported a significant study using the FE program ABAQUS in which they investigated the effect of footing width and relative density (D_r) on the mobilised bearing resistance. The CPT q_c values used to normalise the footing resistance were derived from a different program CONPOINT (Salgado and Randolph, 2001). Their analyses shown in Figure 3 indicate that α decreased when the relative density of the soil increased and the footing width reduced. The rate at which α increased with the footing width depended on the relative density of the soil, with an increase of 35% being noted for $D_r = 90\%$ when the footing width increased from 1 m to 3 m, whilst the increase was only 5% for $D_r = 30\%$.

The variability in suggested α values from previous studies arises at least in part due to the limited

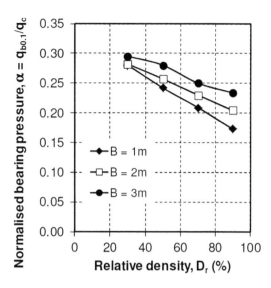

Figure 3. ormalized Base Resistance $q_{b0.1}/q_c$ versus Relative Density D_R; (after Lee and Salgado 2005).

database of footing tests from which to investigate factors which might influence the footing response. Large-scale footing tests are expensive and time consuming, and therefore, most field tests consider a relatively limited range of either footing width and/or depths in sand where the relative density is relatively constant. Gavin et al. (2009) compiled data from 22 field experiments on model and full-scale footings at five test sites. Data from the load tests performed by Briaud and Gibbens (1999) in Texas, Ismael (1985) in Kuwait, Anderson et al. (2007) at Green Cove, Gavin et al. (2009) from Blessington and Lehane et al. 2008, from Shenton Park were collated. The footing width varied from 0.1 m to 3 m, the q_c value (average over ±1.5B) was in the range 3,330 to 14,500 kPa, and the foundation depth varied from 0.1 m to 2 m below ground level, bgl. The normalised pressure-settlement response is set out in Figure 4a.

Whilst many of the footing tests did not reach the normalised displacement level of 10%, the trend is for all sites to converge as s/B increases and tend towards α of approximately 0.2 at s/B = 10%. However, the initial stiffness response at the five sites varied considerably, and the data is recast in Figure 4b in order to examine the rate of mobilisation of the bearing resistance. The data confirms that the initial pressure-settlement response is site dependent, with a tendency for the normalised resistance to mobilise most quickly for the tests in Texas and most slowly at Shenton Park. Also included in Figure 4b are foundation settlement measurements from 23 buildings as collated by Papodopoulos (1992). The measurements relate

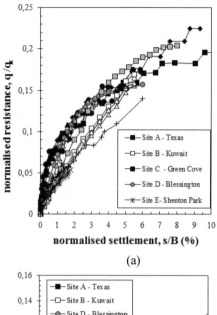

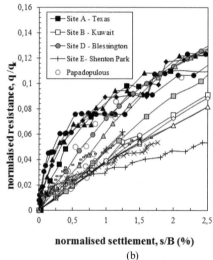

Figure 4. (a)Pressure settlement response from five test sites (b) Initial response of footing tests (after Gavin et al. 2009).

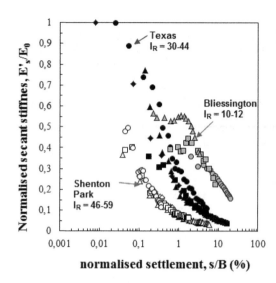

Figure 5. Variation of normalised secant stiffness with normalised settlement.

to foundations with B between 0.5 and 21.7 m, founded in sand where q_c ranged from 1800 to 19,000 kPa. Despite the range of parameters considered, the discrete measurements of maximum settlement are bounded by the experimental data presented in Figure 4b.

To gain further insight into the mobilisation rate, the rate of degradation of the secant stiffness measured at four sites where small strain stiffness data was available are compared in Figure 5. It is of interest to compare the normalised settlement level at which the secant stiffness reduces to 20% of the small strain Young's modulus ($E'_s/E_0 = 0.2$). This occurs at s/B = 0.2%, 1% and 10% at Shenton Park, Texas and Blessington respectively. The stiffness degradation rate appears to be related to the rigidity index, I_R (ratio of small strain shear modulus, G_0 to soil strength, q_c) with high I_R values resulting in rapid stiffness degradation.

The relatively low applied pressure imposed by most regular structures ensures that the factor of safety against ultimate failure of these foundations is high. As a result of this, the focus of designers is often to accurately predict settlement under working stresses. Settlement prediction models that require an estimate of the secant stiffness E'_s of sand are in common use. A number of direct correlations between E'_s and q_c have been proposed:

$$E'_s = \beta\, q_c \qquad (3)$$

Das (1983) compiled β values recommended by 13 sources that ranged from 1 to 3. However, Lehane et al. (2008) and others state that at relatively small strain levels there is a weak dependence between E'_s and q_c, with the effects of stress level and ageing being dominant. Given that many foundations are designed for a maximum settlement often of the order of 25 mm, the resulting normalised settlement beneath foundations of width between 1 m and 10 m, results in s/B% in the range 0.25–2.5%. The variation of β values mobilised from the field tests described above, and shown in Figure 6, shows a very large range of operational β values at these normalised settlement levels, confirming that constant β values should be used with caution. It is worth noting that at the settlement levels typical for full-scale foundations, β values are typically higher than 3.

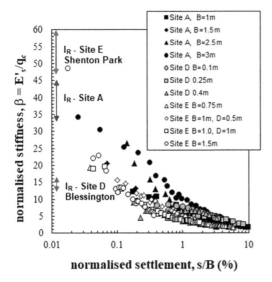

Figure 6. Variation of β (= E'_s/q_c) with normalised settlement.

Table 1. Recommendations for bearing capacity factors and limiting values for offshore piles (API 2007).

Density	N_q	q_b limit (MPa)
Loose	–	
Medium	20	5
Dense	40	10
Very Dense	50	12

where N_q is a bearing capacity factor and σ'_v is the vertical effective stress at the pile base. N_q factors typically vary with ϕ' or sand density as shown in Table 1. Another common feature is the adoption of limiting upper-bound values. Randolph and Gouvernec (2011) note that these limiting values are included as field data suggests that N_q reduces with increasing stress level.

The N_q values recommended for bored and cast in-place piles are typically assumed to be 50% of those derived for driven piles.

2.3 Summary

Due to the difficulty of sampling and testing the strength and stiffness properties of sand, direct correlation between the ultimate bearing capacity and settlement response of footings on sand and in-situ testing such as CPT offers a viable alternative that appears to give consistent results.

Data from field tests presented by Gavin et al (2009) suggest that the normalised bearing pressure normalised by CPT q_c, mobilised at a footing settlement of 10% of the footing diameter, α is 0.2. This agrees reasonably well with the findings from a larger database study by Mayne et al. (2012) that considered a wider range of soil types and foundation size and concluded that a constant α value of 0.18 was appropriate. Both studies propose simple non-linear elastic models that allow the complete pressure-settlement response of the foundation to derived thus providing important insights into the operation behaviour of footings, and particularly the relatively large stiffness response at typical operation pressure levels.

3 BASE RESISTANCE OF PILES

3.1 Background

In many design guides and codes the end bearing resistance (q_b) mobilised by a deep footing in sand is calculated using a reduced form of Equation 1:

$$q_b = N_q \sigma'_v \quad (4)$$

3.2 Closed-ended driven piles

Because of the similarities between the penetration mechanisms and the geometry of closed-ended piles and the CPT, a number of empirical correlations between the CPT end resistance q_c and q_b have been proposed. Recent design methods for driven closed-ended or full-displacement piles in sand have been shown to have a relatively high reliability (Chow 1997, Gavin 1999 and Lehane et al. 2005) and have been widely accepted in industry. These techniques generally estimate the base resistance at relatively large pile base settlement (s_b), typically at 10% of the pile diameter, $q_{b0.1}$, using an empirical reduction factor α_p:

$$q_{b0.1} = \alpha_p q_c \quad (5)$$

Based on a database study Jardine et al. (2005) suggest that α_p reduced from 0.63 to 0.43 as the pile diameter increased from 200 mm to 500 mm, Lehane et al. (2005) found that an α_p value of 0.6 gave the best-fit to a database of instrumented pile load tests with diameters ranging from 0.2 m to 0.68 m. Randolph (2003) and White and Bolton (2005) argued that once appropriate averaging techniques were adopted to derive design q_c values and the effects of residual loads were accounted for, a constant α_p factor can be adopted which is independent of pile diameter and tended towards $q_b = q_c$ at large pile base displacements.

3.3 Open-ended driven piles

For partial displacement (open-ended) pipe piles, model and full-scale pile tests reported by Lehane and Gavin (2001) and Foye et al. (2009) show that direct correlations between α_p (based on the average pressure mobilised over the entire pile base area) and q_c which are independent of pile diameter or sand state can be determined once the degree of plugging during pile installation are included. The plugging behaviour is best quantified through the incremental filling ratio (IFR), which compares the rate of soil intrusion during pile installation with IFR = 1 for a fully coring pile (which causes minimal disruption) and IFR = 0 for a pile with a fully formed plug, which prevents soil intrusion and results in what is effectively a closed-ended pile.

However, in reality, plug development during installation is rarely recorded and the final plug length ratio, PLR = soil plug length/pile embedment depth is only known. In the field, plugging appears to be controlled by pile diameter, sand density and installation method. The majority of large diameter (D >500 mm), piles driven offshore appears to have PLR values close to one. Significant plugging seems to occur only in the case of smaller diameter piles, or piles installed by jacking.

Gavin and Lehane (2005) present data shown in Figure 7 of plug stress mobilised during jacking strokes (closed-symbols) and static load tests (open-symbols) for pipe piles in jacked into loose sand. The normalised plug stress varies linearly with IFR from a minimum value in the range 0.15 to 0.2 that is in keeping with values suggested by Lehane and Randolph (2002) from numerical analyses of fully-coring piles. As IFR reduced, q_{plug}/q_c increased linearly. The trend was shown to be consistent over a range of installation methods (driving and jacking) and sand sate (Gavin and Lehane 2003).

Lehane and Gavin (2001) report measurements of the annular stress mobilised during installation of open-ended piles that were independent of IFR (with $q_{ann} = q_c$). Lehane et al. (2005) propose that the average base stress (combining the plug and annular components) mobilised for an open ended pile be calculated using Equation 6:

$$q_{b0.1} = 0.15 + 045\, A_{r,eff} \tag{6}$$

where $A_{r,eff} = 1 - FFR\,(D_i/D)^2$, and FFR is the IFR value at the end of driving.

Equation 6 suggests that α_p increases from ≈ 0.15 for a fully coring pile (IFR = 1) to 0.6 for a fully-plugged pile.

3.4 Partial displacement screw injection piles

Funderingstechnieken Verstraeten bv. performed three static compression load tests on screw injection piles installed in dense sand at a site in Terhausen, Netherlands. The piles which had shaft diameters of 0.46 m, base diameters of 0.56 m and embedded lengths of between 20.2 m and 20.3 m were instrumented with strain gauges along their length. The purpose of the test was to assess recent changes in pile capacity factors introduced in the Dutch standard NEN-EN 9997-1 introduced on January 1st 2017 that reduced the α_p values for all piles by 30%. The result was that for screw injection piles the value decreased from 0.9 to 0.63.

Estimates of the pile capacity at the test site were made using the pre-2017 design value and the piles were load tested to this capacity in an attempt to validate the old design approach. The CPT profiles at the test site are shown in Figure 8. The estimated pile capacity ranged from 5750 to 6100 kN. The load test on pile 1 was terminated an applied load of 5874 kN when the pile head settlement reached 23 mm (i.e. less than 5% of the pile diameter). The other piles were loaded up to failure at ultimate loads of 6096 kN (Pile 3) and 6312 kN (Pile 5) causing pile displacements of 60 mm.

The base pressure-settlement response of the piles is illustrated in Figure 9, showing that the initial stiffness response of the three piles was similar. Pile 1 mobilised the highest resistance, 13.7 MPa despite not reaching failure. Pile 3 mobilised an ultimate base resistance of 12 MPa and Pile 5 had the lowest resistance of 10 MPa.

The back-figured α_p value depends on the design average q_c value adopted. A comparison of values derived from the Dutch averaging method, where q_c is averaged over a distance -8B to + 4B and assuming q_c averaged over a distance ± 1.5B above the pile base are shown in Table 2. Adopting the Dutch averaging technique the measured α_p is 1.12, 80% higher than the current value (of 0.63). Using the ± 1.5B suggests that the ultimate end bearing resistance of these piles approximates q_c.

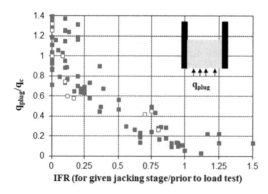

Figure 7. Effect of Incremental Filling Ratio on the plug stress mobilised by open-ended piles in sand (after Gavin and Lehane 2005).

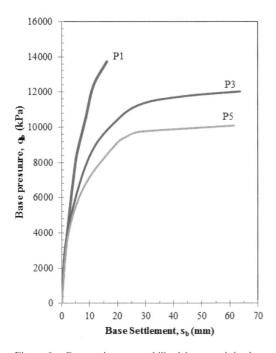

Figure 8. CPT profile at Terhausen test site.

Figure 9. Base resistance mobilised by gout injection piles at Terhausen test site.

Table 2. Comparison of base capacity factors back-figured from Terhausen test site using different CPT averaging techniques.

Pile	q_c Dutch (MPa)	$\alpha_{p;Dutch}$	±1.5B	$\alpha_{p;Dutch}$
Pile 1	9.75	1.40	12	1.23
Pile 3	10.90	1.10	15	0.80
Pile 5	11.60	0.86	11.8	0.85
Average		1.12		0.96

3.5 Bored piles

The earliest correlation between the end bearing resistance of bored piles and q_c were suggested by (Meyerhof 1956; Meyerhof 1976; Meyerhof 1983). Based on theoretical and experimental studies on deep foundations, he introduced two reduction factors, C_1 for scale effects and C_2 to account for shallow penetration into dense strata to predict the ultimate end bearing resistance (q_{bu}) of a bored pile:

$$q_{bu} = 0.3 \, q_c \, C_1 \, C_2 \qquad (7)$$

The LCPC or French method (Bustamante and Gianeselli 1982) was developed from a database of 197 pile tests on driven and bored piles in a range of soil conditions. An α_p value of 0.2 is recommended for bored piles in sand and gravel. The design q_c is the average from ±1.5B having excluded values that are in excess of ± 30% of the average. De Cock et al. (2003) compiled a review of pile design practice in Europe and found that α commonly ranged between 0.15 and 0.2, and typically did not depend on the diameter or the length of the pile. Eurocode 7, Part 2, suggests that α_p values, which although independent of footing width and depth, reduce from 0.2 for q_c values up to 15 MPa, to 0.16 for q_c = 20 MPa. Lehane (2008) reported a database study of Continuous Flight Auger (CFA) piles where the pile length varied from 4 m to 10.5 m. The database contained both straight and expanded base piles and recorded α_p values that increased from approximately 0.15 for 4 m long piles, to approximately 0.4 for a 10.5 m long pile.

Gavin et al. (2013) compiled a pile test database comprised of 20 static maintained load tests performed on non-displacement piles installed in sand where the piles were loaded to settlements in excess of 10% of the pile diameter. The diameter B of the piles ranged from 0.2 m to 1.5 m, while their length L ranged from 4 m to 26.5 m, with L/B in the range 4 to 37. They were founded in sand where the CPT end resistance, $q_{c1.5B}$ value ranged between 2 MPa and 40 MPa.

In the assessment of mobilised α_p values, the design $q_{c(1.5B)}$ was adopted. The variation of α_p with pile diameter, length, relative density and CPT q_c value is shown in Figure 10. Although there is considerable scatter in the results, there is no suggestion that α_p varied in a consistent manner with any of the parameters considered in the assessment, with an average α value for the database piles of 0.24.

Tolooiyan and Gavin (2013) performed finite element analysis using PLAXIS to investigate the factors affecting α_p for bored piles in sand. The sand was modelled using the Hardening Soil (HS) model described by Schanz et al. (1999) with the HS model parameters calibrated from lab-test data. Cavity expansion analyses were performed using a procedure described by Xu and Lehane (2008) and Tolooiyan and Gavin (2011) in order to predict the CPT q_c profile for Blessington sand. The FE generated values are compared in Figure 11 with a four typical q_c profile measured at Blessington. It is clear that the FE model provided a reasonably good albeit lower-bound estimate of the measured CPT q_c profile.

The effect of pile width, B on the bearing pressure mobilised by piles in Blessington sand was considered by analysing a pile of fixed length of 6 m with a diameter ranging from 0.2 m to 0.8 m. The piles were loaded until the settlement reached 10% of the pile diameter, see Figure 12a. The maximum end bearing resistance ($\approx$5,500 kPa) of all piles was similar. However, the settlement required to achieve this resistance increased in proportion to the pile diameter. The normalised pressure-settlement response is shown in Figure 12b. This reveals that the normalised stiffness response for the piles was very similar, with $\alpha = 0.31$.

Considering the effect of pile length in Figure 13, a tendency for a slight reduction in α_p

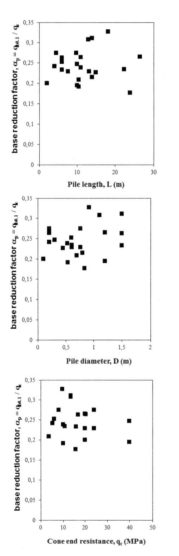

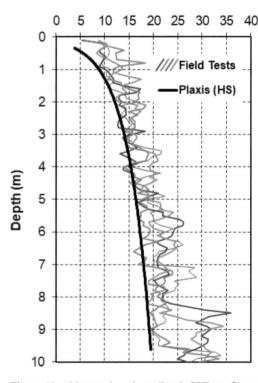

Figure 10. α_p values back-figured from load test database (a) effect of pile length, diameter, and strength.

Figure 11. Measured and predicted CPT profile at Blessington.

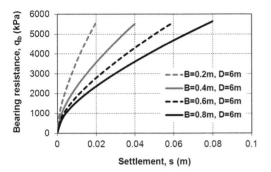

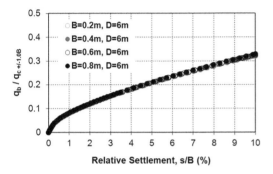

Figure 12. Bearing pressure mobilised by piles installed in Blessington (a) effect of pile width; (b) normalised pressure-settlement response.

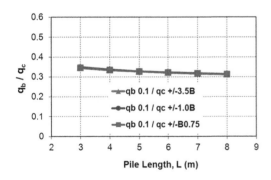

Figure 13. Effect of pile length L on normalised bearing pressure.

with depth is noted with the value decreasing from 0.34 to 0.31 as the pile length increased from 3 m to 8 m. The influence of CPT averaging technique (from ± 0.75B to ± 3.5B) technique is also considered. However, for Blessington sand the effect is negligible.

In order to investigate the effect of sand density on the mobilised α factor, sensitivity analyses were performed using three sand deposits. These were Tanta sand from Egypt which has an in-situ relative density D_r = 75% (reported by El Sawwaf (2005); El Sawwaf (2009)), Monterey sand from the United States with D_r = 65% (Wu et al. (2004) and Yang et al. (2008)), and Hokksund sand from Norway with D_r = 50% (Tefera et al. 2006). Synthetic CPT q_c profiles were derived for these deposits using the procedure as for Blessington sand. The predicted CPT q_c profiles are illustrated in Figure 14.

The sensitivity analyses were performed using a fixed pile diameter (B = 0.4 m) and a pile length which varied from 4 m to 8 m. The pressure-settlement values predicted are shown in Figure 15. The highest q_b values were measured in Tanta sand and the lowest were in Hokksund sand. The q_b values at all sites increased as the pile length increased from 4 m to 8 m. The increase was in the range 56% to 77%, with the highest increase being in Tanta sand.

When the bearing pressure mobilised at a pile base settlement of 10% was normalised by the CPT q_c value averaged over ± 1B, $α_p$ values in the range 0.32 to 0.33 were determined. There were clear variations in the rate of mobilization at low settlement levels across the sites, with the normalised resistance developing more slowly as the relative density increased.

The influence of CPT averaging technique and pile length on $α_p$ is considered in Figure 16. Consistent

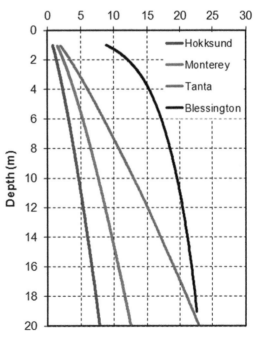

Figure 14. FE derived CPT q_c profile for Hokksund, Monterey and Tanta sand.

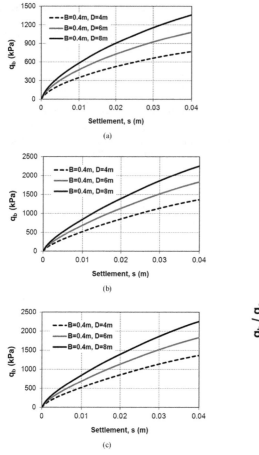

Figure 15. Pressure-settlement values predicted for: (a) Hokksund; (b) Monterey; (c) Tanta sand.

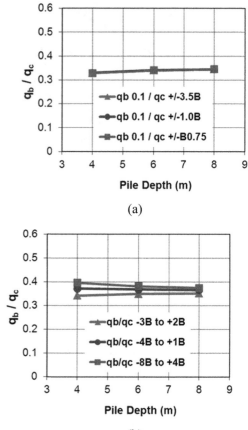

Figure 16. Effect of CPT averaging technique and pile length on α_p values in Hokksund Sand.

trends were found for all sites where α_p was reasonably constant when q_c was averaged over equal distances above and below the pile base, see Figure 16a. When q_c was averaged with a bias for values either above or below the pile base, See Figure 16b, higher variability and a more pronounced length bias was observed with the highest α_p values being inferred when using the Dutch average technique.

On the basis of these analyses, it appears that an approximate constant α_p factor of 0.3 would produce reasonable estimates of the end bearing resistance of bored piles in sand. This value is 50% higher than values typically used in practice and 20% higher than the α_p value of 0.24 inferred from the database study. To investigate one possible reason for the difference between the field response and FE analyses, it is of interest to consider the effect of loading rate on the mobilisation of α_p. Gavin et al. (2013) describe load tests performed on instrumented continuous flight auger piles installed in dense sand in Killarney, South-West Ireland. Load tests were performed on two piles, a 450 mm diameter, 15 m long pile and an 800 mm diameter, 14 m long pile. The load test procedure involved a maintained load test (MLT) followed by a fast-loading, constant rate of penetration (CRP) test. The results of the MLT portion of the load test are shown in Figure 17a where it is clearly evident that when the applied base pressure exceeded 1500 kPa, the piles experienced creep during load increments. When the normalised base displacement reached 10% of the pile diameter, the α_p factor approached 0.24.

When the piles were reloaded in the CRP test (See Figure 17b), significantly higher base resistance was mobilised. Whilst the loading history would affect the initial pressure-settlement response, the α_p factors mobilised in the fast loading tests (where

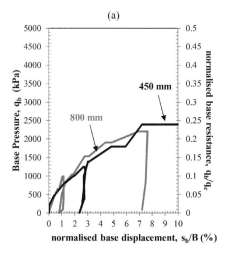

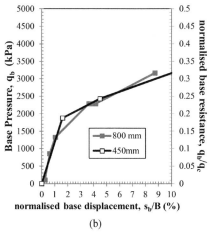

Figure 17. Effect of loading rate or test method on α_p mobilised by CFA piles in sand.

creep effects were minimised) exceeded 0.31. The soil models used in the numerical analyses did not model sand creep which clearly affected the MLT result and thus the mobilised base resistance.

4 SHAFT RESISTANCE OF PILES

4.1 Background

The peak unit shaft resistance (τ_f) mobilised by a pile in sand can be estimated using earth pressure theory as:

$$\tau_f = K \sigma'_v \tan\delta_f \qquad (8)$$

where K is the earth pressure coefficient, σ'_v is the in-situ vertical effective stress, and δ_f is the soil-pile interface friction angle. As with all earth pressure approaches, the difficulty with the application of Equation 8 is in the choice of an appropriate K value for design. For bored piles, Reese and O'Neill (1999) suggest K/K_0 (where K_0, is the coefficient of earth pressure at rest), varies with the pile construction method, varying from 0.67 for a pile excavated using slurry, to 1.0 for a pile formed in a dry excavation. K_0 is notoriously difficult to measure, but may be estimated using the method proposed by Kulhawy and Mayne (1990):

$$K_0 = 1 - \sin\phi' - \text{Normally Consolidated} \qquad (9)$$
$$K_0 = (1 - \sin\phi') \text{OCR}^{\sin\phi'} - \text{Over Consolidated}$$

where ϕ' is the friction angle and OCR is the Over-Consolidation Ratio.

For displacement piles, K will be changed during pile installation due to large stress changes as the pile base moves through the penetration depth and friction fatigue occurs during cyclic installation. Values in the range K = 1–2 are often used in practice. Due to uncertainties regarding input parameters such as ϕ', OCR and δ_f, and the effect of installation method, the use of CPT based design methods cone Penetration Test (CPT) are increasing:

$$\tau_f = \alpha_s q_c \qquad (10)$$

A range of α_s values that depend on pile type are recommended in many design codes, with values typically being lowest for bored piles and highest for displacement piles. For example, in the Netherlands NEN 9997-1-2016 recommends α_s value that range from 0.06 for bored or driven open-tube piles, to 0.01 driven concrete or closed-ended steel tubes and up to a maximum of 0.014 for a driven cast-in-place piles. In Belgian practice, summarized by Huybrechts et al. (2016), α_s depends on q_c, pile type and roughness. A limiting maximum shaft resistance of 150 kPa is imposed for q_c values > 20 MPa.

Intensive research on displacement piles reported in Jardine et al. (2005) and Lehane et al. (2005) show that the local shaft resistance is given by:

$$\tau_f = (\sigma'_{hc} + \Delta\sigma'_{hd}) \tan \delta_f \qquad (11)$$

where: σ'_{hc}, is the fully equalized horizontal effective stress after pile installation and $\Delta\sigma'_{hd}$ is a component derived by dilation during loading.

Chow (1997) examined profiles of σ'_{hc} recorded by an instrumented pile installed at two sites and found that σ'_{hc} values at a given location on the pile were almost directly proportional to the q_c value at that level and the distance from the level to the pile base (h) normalised by the pile diameter (D). These findings were incorporated into the widely used design method for displacement piles known as the Imperial College (IC-05) design method (Jardine et al. 2005) and a similar approach known as the University of Western Australia (UWA) method (Lehane et al. 2005), where:

$$\sigma'_{hc} = 0.03 \, q_c \, (h/D)^{-0.5} \qquad (12)$$

A minimum h/D value of 2 should be used is Equation 12.

Equation 12 suggests that α_s is highest near the pile tip and reduces with increasing distance from the pile tip. Lehane (1992) suggests that the dilation induced increase in horizontal stress ($\Delta\sigma'_{hd}$) could be predicted using cavity expansion theory:

$$\Delta\sigma'_{hd} = \frac{4 G \, \delta_h}{D} \qquad (13)$$

where G is the shear modulus of the soil mass and δ_h is the horizontal displacement of a soil particle at the pile-soil interface.

The IC-05 and UWA design approaches have been shown to provide more reliable estimates of the shaft capacity of piles and accurate predictions of the distribution of mobilised local shear stress on closed-ended displacement piles by Chow (1997), Gavin (1998), Schneider (2007) and others. Given the prevalence of driven open-tube piles, particularly in the offshore sector, both the IC-05 and UWA methods allow for a reduction of shaft stress due to the lower displacement resulting from installation of these piles. Gavin et al. (2011) note that whilst the two approaches give very similar predictions for shaft capacity of closed-ended piles, differences in how they address the issue of plugging can result in significantly different estimates for open-ended piles.

4.2 Driven cast-in-place piles

Flynn and McCabe (2016) describe instrumented pile load tests performed on 3 driven cast-in-place piles installed at a site near Coventry, in the United Kingdom. The piles that were formed by driving a 0.32 m diameter hollow steel tube with a sacrificial circular steel plate at the base were instrumented with strain gauges at four levels. When they reached their final penetration depths of between 5.5 m and 7 m bgl. the steel tube was filled with concrete and the steel casing was withdrawn. The CPT profile at the site is shown in Figure 18. The ground conditions comprise made ground and stiff sandy clay to approximately 1.8 m bgl. underlain by medium dense to dense sand to a depth of ~ 6.5 m. Below this depth the sand becomes increasingly gravelly.

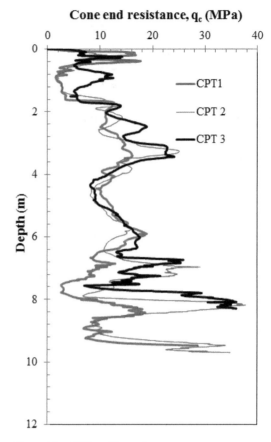

Figure 18. CPT profiles at Coventry.

The piles were load tested in compression between 19 and 23 days after installation.

The mobilisation of average shaft resistance (τ_{av} = total shaft resistance/shaft area) during the load tests is shown in Figure 19. The three piles exhibited similar initial stiffness response with peak α_s values in the range 0.094 to 0.14.

A noticeable feature of the response is that relatively large normalised dis-placement was required to mobilise the peak capacity (between 3% and 13% of the pile diameter) and in the case of Piles 1 and 3, the resistance was still increasing up to the end of the load test.

4.3 Continuous Flight Auger (CFA) Piles

Gavin et al. (2009) presents the results of instrumented load tests performed to study the development of shaft resistance on Continuous Flight Auger piles installed in sand. Ground conditions at the test site consist of approximately 2 m of mixed (sand, silt and clay) deposits overlying a deep deposit of sand. The CPT end resistance, q_c

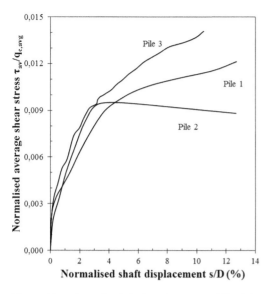

Figure 19. Normalised shaft resistance mobilised by driven cast-in-place piles (after Flynn and McCabe 2016).

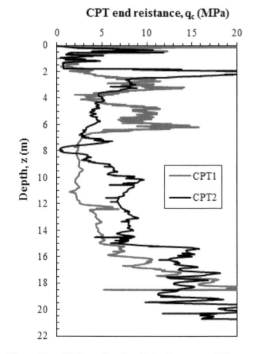

Figure 20. CPT profiles for CFA pile tests at Killarney (after Gavin et al. 2009).

measured in the vicinity of the test piles are shown in Figure 20.

Two instrumented test piles were installed at the site. The first pile was installed using an 800 mm

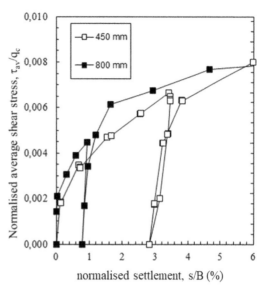

Figure 21. Normalised shaft resistance mobilised by Continuous Flight Auger Piles (after Gavin et al. 2013).

diameter auger to 14 m bgl., and the second using a 450 mm auger to 15 m bgl. The piles were instrumented with strain gauges at four levels in order to separate base and shaft resistance components and to determine the distribution of shaft resistance along the pile.

Static load tests in compression were performed and the average shaft resistance (τ_{av}) mobilised during the static load tests of between 35 and 36 kPa was almost identical on both the 450 mm and 800 mm diameter piles suggesting that scale effects were insignificant. The resulting α_s ($= \tau_{av}/q_{cav}$) value of 0.008, shown in Figure 21 are similar to those used for the design of displacement piles in sand. The relatively large displacement required to mobilise the peak resistance is again a feature of the pile tests.

4.4 Screw injection piles

The average shaft resistance mobilised by the screw injection piles installed in dense sand at a site in Terhausen, Netherlands, described in Section 2.4 is shown in Figure 22a. The initial stiffness response of the piles which had shaft diameters of 0.46 m were remarkably similar. The load test on Pile 1 was stopped before it reached peak resistance as the pile mobilised a much higher base resistance than the other test piles, see Figure 9.

Notwithstanding this, it would seem that the ultimate shaft resistance of the piles was in the range 110 kPa to 130 kPa, However, displacements in excess of 10% of the pile shaft diam-

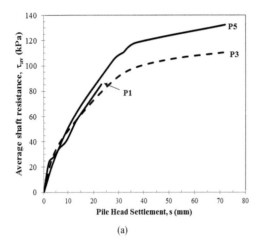

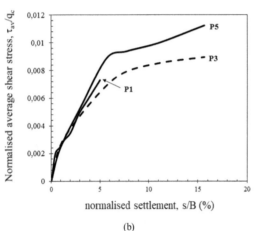

Figure 22. Shaft resistance (a) and normalised resistance mobilised (b) mobilised by Screw Injection Piles at Terhausen.

eter were required to mobilise this resistance. The back-figured α_s values shown in Figure 22b show that the Dutch standard NEN-EN 9997-1 recommended α_s value of 0.09 would provide a reasonable estimate of the fully mobilised shaft resistance for these piles.

4.5 Distribution of normalised shaft resistance on piles in sand

Whilst the α_s values mobilised by the test piles described above conformed broadly with the constant values proposed in many codes and seemed to depend on pile type, with higher values generally pertaining to driven piles, there remains some variation in α_s values proposed by different codes. Some insight into possible reasons for this can be determined by the instrumented load tests. The strain gauges on the piles allowed the local shear stress profiles to be determined.

The normalised local shear stress profile along the driven cast-in-place (DCIP) pile (Flynn and McCabe 2016) is compared in Figure 23 to the profile predicted using the UWA-05 approach with

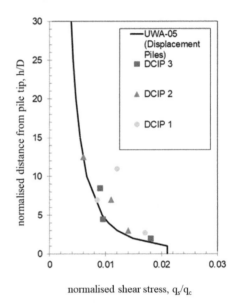

Figure 23. Comparison of distribution of shear stress on driven cast-in-place pile with prediction of UWA-05 method.

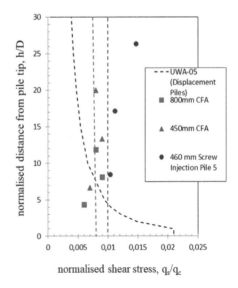

Figure 24. Comparison of distribution of normalised shear stress on CFA and Screw-Injection piles with prediction of UWA-05 method.

$\delta_f = \phi'_{cv}$). It is clear that a method which includes for the effects of friction fatigue provides a very good match to the measured response and suggests that for piles where friction fatigue occurs during installation that α_s should vary with pile geometry.

The normalised local shear stress values for the CFA piles and a typical screw injection pile are shown in Figure 24. For these piles that do not experience friction fatigue during installation, a constant α_s value is suitable to describe the shaft resistance. The value is lower than that applying to driven piles over a distance of 5D from the pile tip. Above this level the piles developed larger α_s values. Thus suggesting that non-displacement piles with deep embedment lengths could mobilise higher average α_s values than piles driven in the same deposit.

5 CONCLUSIONS

Whilst conventional earth pressure approaches remain in common use for the design of shallow and deep foundations in sand, researchers have highlighted a number of challenges related to the applicability of these methods. In all cases, obtaining high quality samples and determining operational frictional angles over the range of stress and strain relevant for design remains an issue. For over-consolidated deposits, accurately determining the earth pressure coefficient is essential and non-trivial. For shallow foundations, issues related to determining bearing capacity factors to address depth and width effects particularly for soil deposits where the strength does not increase linearly with depth have been highlighted. For pile design, the impact of installation effects; for example friction fatigue which causes complex changes to in-situ horizontal stress regime around a pile and plugging effects that results in large changes to the strength and stiffness of the sand below the pile base have been quantified in recent field tests.

Given that the CPT provides a continues indirect measurement of the strength and stiffness properties over the complete range of sand state, correlations between the cone end resistance q_c and foundation behaviour are in common use and have been embedded in the design codes in many countries. However, many of these codes give conflicting guidance on design. In this paper, results from lab and field experiments on instrumented piles and finite element analyses are used to explain the physical basis for the regional variations in CPT based approaches.

For the design of shallow foundations a number of workers have proposed α values that vary with sand state and footing dimensions. Field tests performed by Briaud (2007), Gavin et al. (2009), Mayne et al. (2012) and others suggest that unique site specific normalised resistance curves can be derived which are independent of footing geometry. A database of field tests collated in this paper suggests that at large normalised settlement levels (s/B = 5–10%), these normalised resistance curves converge and are independent of sand state. However, the initial stiffness response of the foundation is strongly dependent on the stress history and knowledge of both the strength (q_c) and small-strain stiffness parameters are crucial for design.

Considering the base resistance mobilised by piles in sand, the CPT q_c value appears to be an ideal design tool. Whilst at very large pile displacements the base resistance, q_b tends to the q_c value, see White and Bolton (2001) and Randolph (2005). The exact displacement required depends on the pile installation method and in some cases could be several pile diameters. At displacement levels typically considered in practice as representing failure, e.g. s/B = 10%, the stiffness response depends on the pile installation method. For low displacement pile types α_p (= $q_{b0.1}/q_c$) in the range of 0.15–0.2 are typically adopted in practice. For displacement piles, q_b/q_c of 0.6 is recommended for closed-ended driven steel and concrete piles. A particular feature of field tests on shallow foundations and bored pile (i.e. where pre-stressing during installation did not occur) was that significant creep occurred during maintained load tests at high stress levels. Numerical analyses of these piles using soil models that do consider creep, showed α_p values around 50% higher than those measured in maintained load tests. In quick load tests where creep was minimised, α_p values similar to those in the finite element model were mobilised. Based on these results, if effects such as loading rate, the definition of failure and residual loads are considered then higher α_p values can be adopted in design codes.

When considering shaft resistance, a number of codes suggest constant α_s (= τ_{av}/q_{cav}) values, with recommended values for displacement piles typically being higher than non-displacement piles. The case histories presented herein strongly suggest that α_s values for driven piles should incorporate a friction fatigue parameter. The absence of friction fatigue effects for non-displacement piles could result in some slender driven piles developing lower α_s than a non-displacement pile installed in the same soil deposit.

ACKNOWLEDGEMENTS

Much of the work presented in this paper was completed together with a number of former research students and collaborators who include: Prof. Barry Lahane, Dr. Ali Tolooiyan, David Cadogan,

Dr. Cormac Reale, Dr. Abidemi Adekunte and others.

The permission of Funderingstechnieken Verstraeten bv. to publish the pile load test data from Terhausen is gratefully acknowledged.

REFERENCES

Anderson, J.B, Townsend, F.C. and Rahelison, L. (2007) Load Testing and Settlement Prediction of Shallow Foundation Journal of Geotechnical and Geoenvironmental Engineering, Vol. 133, No. 12, pp 1494–1502.

API (2007) Recommended Practice for Planning, Designing and Constructing Fixed Offshore Platforms, 22nd edn. American Petroleum Institute, Washington, DC, USA, API RP2 A.

Briaud, J.L. (2007), 'Spread Footings in Sand: Load Settlement Curve Approach', *Journal of Geotechnical and Geoenvironmental Engineering, ASCE*, vol. 133, no. 8, pp. 905–920.

Briaud, J.L. & Gibbens, R. 1999. Behaviour of five large spread footings in sand. Journal of geotechnical and geoenvironmental engineering 125(9): 787–796.

Bustamante, M. & Gianeselli, L. (1982), 'Pile Bearing Capacity Predictions by Means of Static Penetrometer CPT', *Second European Symposium on Penetration Testing, ESOPT-II*, Amsterdam, The Netherlands, pp. 493–500.

Chow, F. (1997). "Investigations into the behavior of displacement piles for offshore structures." PhD Thesis, Uni-versity of London (Imperial College), London.

Das B.M. (1983) Advanced Soil Mechanics. McGraw-Hill, New York.

De Cock, F., Legrand, C. & Huybrechts, N. (2003), 'Overview of design methods of axially loaded piles in Europe', *8th European Conf. on Soil Mech. and Geotech. Eng. Report of ERTC3-Piles, ISSMGE Subcommittee*, Prague, pp. 663–715.

El Sawwaf, M. (2005), 'Strip footing behaviour on pile and sheet pile-stabilized sand slope', *Journal of Geotechnical and Geoenvironmental Engineering, ASCE*, vol. 131, no. 6, pp. 705–715.

El Sawwaf, M. (2009), 'Experimental and numerical study of eccentrically loaded strip footings resting on reinforced sand', *Journal of Geotechnical and Geoenvironmental Engineering, ASCE*, vol. 135, no. 10, pp. 1509–1518.

Eslaamizaad, S and Robertson,P.K. Cone penetration testing tests to evaluate bearing capacity of foundations in sands. Proceedings of the 49th Conference of the Canadian Geotechnical Society, 1986, 429–438.

Foye, K.C., Abou-Jaoude, G., Prezzi, M., and Salgado, R. (2009). Resistance factors for use in load and resistance factor design of driven pipe piles in sands. Journal.of Geotechnical. Geoenvironmental. Eng., 135(1), 1–13.

Flynn, K.N. and McCabe, B.M. (2016), Shaft resistance of driven cast-in-situ piles in sand, Canadian Geotechnical Journal, Vol. 53, Issue 1, pp 49–59.

Gavin, K. (1999) Experimental investigation of open and closed ended piles in sand', [thesis], Trinity College (Dublin, Ireland). Department of Civil, Structural and Environ-mental Engineering, pp 369

Gavin, K., Adekunte, A. & O'Kelly, B. (2009), 'A field investigation of vertical footing response on sand', Proceedings of the Institution of Civil Engineers-Geotechnical Engineering, vol. 162, no. 5, pp. 257–267.

Gavin, K.G, and Lehane (2003), The Shaft capacity of pipe piles in sand The Shaft capacity of pipe piles in sand, Ca-nadian Geotechnical Journal, Vol. 40, No.1, pp36–45, (2003), doi:10.1139/t02–093

Gavin, K. & Lehane, B. (2005), Estimating the end bearing resistance of pipe piles in sand using the final filling ratio. Proceedings of the International Symposium on Frontiers in Offshore Geotechnics. ed. / M.J. Cassidy; S. Gourvenec. Vol. n/a Perth, Australia. ed. The Netherlands: CRC Press/Balkema, 2005. p. 717–723.

Gavin., K., Cadogan., D. and Casey., P (2009)The shaft capacity of CFA piles in Sand ASCE Journal of Geotech-nical and Geoenvironmental Engineering, Vol 135, Issue 6, pp 790–798.

Gavin., K., Cadogan., D., Casey., P. & Tolooiyan., A. (2013), 'The Base Resistance of Non-Dislacement Piles in Sand –: Field Tests', *ICE Journal of Geotechnical Engineering*, Vol. 166, 6, pp 540–548

Houlsby, G.T. and Cassidy, M.J.. A plasticity model for the behaviour of footings on sand under combined loading. Géotechnique, 2002, 52, No. 2, 117–129.

Huybrechts, N, De Vos, M, Bottiau, M. and Maertens, L. (2016) Design of piles—Belgian practice, ISSMGE-ETC3, International Symposium on Design of Piles in Europe, Leuven, Belgium, April eds De Vos et al.

Ismael N.F. Allowable pressure from loading tests on Kuwaiti soils. Canadian Geotechnical Journal, 1985, 22, No. 2, 151–157.

Kulhawy, F.H., and Mayne, P.W. 1990. "Manual on estimating soil properties for foundation design." Rep. No. EL-6800, Electric Power Research Institute, Palo Alto, Calif.

Lee, J.H. & Salgado, R. (1999), 'Determination of pile base resistance in sands', *Journal of Geotechnical and Geoenvironmental Engineering, ASCE*, vol. 125, no. 8, pp. 673–683.

Lee, J.H. and Salgado, R. (2005). Estimation of Bearing Capacity of Circular Footings on Sands Based on Cone Penetration Test. Journal of Geotechnical and Geoenvironmental Engineering, Vol. 131, No. 4, April 2005, p. 442–452.

Lehane, B.M. (1992). "Experimental investigations of pile be-havior using instrumented field piles." PhD Thesis, Univer-sity of London (Imperial College), London.

Lehane, B.M. (2008), 'Relationships between axial capacity and CPT qc for bored piles in sand', *5th International Symposium on Deep Foundations on Bored and Auger Piles (BAP V)*, GHENT, BELGIUM, pp. 61–74.

Lehane, B.M and Gavin, K.G., (2001) Experimental inves-tigation of the factors affecting the base resistance of open-ended piles in sand. Journal of Geotechnical and Geoenviromental Engineering, ASCE Vol. 127, No. 6, June (2001). doi:10.1061/(ASCE)1090–0241(2001)127:6(473), p 473–480.

Lehane, B.M. and Randolph, M.F., (2002) Evaluation of a minimum base resistance for driven piles in siliceous sand,. Journal of Geotechnical and Geoenviromental Engineering, ASCE Vol. 128, pp 198–205

Lehane, B.M., Doherty, J.P. & Schneider, J.A. (2008), 'Settlement prediction for footings on sand', *Deformational Characteristics of Geomaterials*, IOS Press/Millpress, Rotterdam, pp. 133–150.

Lehane, B.M., Scheider, J.A. & Xu, X. (2005), *CPT based design of driven piles in sand for offshore structures*, University of Western Australia, Australia.

Mayne, P.W., Uzielli, M. and Illingworth, F. 2012. Shallow footing response on sands using a direct method based on cone penetration tests. Full Scale Testing and Foundation Design(Proceedings GSP 227 honoring Bengt Fellenius), ASCE, Reston, Virginia: 664–679.

Meyerhof, G.G. (1956), 'Penetration tests and bearing capacity of cohesionless soils', *Journal of the Soil Mechanics and Foundations Division*, vol. 82, no. SM1, pp. 1–19.

Meyerhof, G.G. (1976), 'Bearing Capacity and settlement of pile foundations', *Journal of the Geotechnical Engineering Division*, vol. 102, no. GT3, pp. 197–228.

Meyerhof, G.G. (1983), 'Scale effects of pile capacity', *Journal of Geotechnical Engineering*, vol. 108, no. GT3, pp. 195–228.

PLAXIS (2002), *Finite element code for plane strain and axi-symmetric modelling of soil and rock behaviour*, ed.^eds 8.2, Plaxis bv, The Netherlands.

Papadopoulus, B.P. (1992), Settlement of Shallow Foundations on Cohesionless Soils, ASCE Journal of geotechnical Engineering, Vol. 118, No.3 pp 377–393.

Randolph, M.F., Dolwin, J. & Beck, R. (1994), 'Design of drivenpiles in sand', *Geotechnique*, vol. 44, no. 3, pp. 427–448.

Randolph, M.F., and Gouvernec S. (2011) Offshore Geotechnical Engineering, Rouledge Publ.

Randolph, M.F., Janmiolkowski, M.B. and Zdravkovic L. (2004) Load carrying capacity of foundations. Proceedings of Advances in Geotechnical Engineering—The Skempton Conference, London, 1, 207–241.

Reese, L.C., O'Neill, M.W. and Touma, F.T. (1976) Behav-iour of drilled piers under axial loading, Journal of Ge-otechnical Engineering, ASCE, Vol 102, No.5: pp 493–51.

Salgado, R. & Randolph, M.F. (2001), 'Analysis of Cavity Expansion in Sand', *International Journal of Geomechanics*, vol. 1, no. 2, pp. 175–192.

Schanz, T., Vermeer, P.A. & Bonnier, P.G. (1999), 'The hardening soil model: formulation and verification', *Beyond 2000 in computational geotechnics: 10 years of PLAXIS International*, Rotterdam.

Schneider, J.A. (2007) Analysis of piezocone data for displacement pile design, PhD thesis, University of Western Australia

Tefera, T.H., Nordal, S., Grande, L., Sandven, R. & Emdal, A. (2006), 'Ground settlement and wall deformation from a large scale model test on a single strutted sheet pile wall in sand', *International Journal of Physical Modelling in Geotechnics*, vol. 6, no. 2, pp. 1–14.

Tolooiyan, A. & Gavin, K. (2011), 'Modelling the Cone Penetration Test in sand using Cavity Expansion and Arbitrary Lagrangian Eulerian Finite Element Methods', *Computers and Geotechnics*, vol. 38, no. 4, pp. 482–490.

Tolooiyan, A and Gavin, K.G, (2013) The base resistance of non-displacement piles in sand, Part II—Numerical Analyses, ICE Journal of Geotechnical Engineering, Vol 166, 6, pp 549–560.

White, D.J. & Bolton, M.D. (2005), 'Comparing CPT and pile base resistance in sand', *Geotechnical Engineering*, vol. 158, no. GE1, pp. 3–14.

Wu, J., Kammerer, A.M., Riemer, M.F., Seed, R.B. & Pestana, J.M. (2004), 'Laboratory Study of Liquefaction Triggering Criteria', *13th World Conference on Earthquake Engineering*, p. 14.

Xu, X. & Lehane, B.M. (2008), 'Pile and penetrometer end bearing resistance in two-layered soil profiles', *Geotechnique*, vol. 58, no. 3, pp. 187–197.

Yang, K.H., Zornberg, J.G. & Wright, S.G. (2008), *Numerical modeling of narrow MSE walls with extensible reinforcements*, Center for Transportation Research (CTR), Austin, Texas.

CPT'18 papers

Evaluating undrained rigidity index of clays from piezocone data

S.S. Agaiby & P.W. Mayne
Geosystems Group, Civil and Environmental Engineering, Georgia Institute of Technology, Atlanta, USA

ABSTRACT: A review on the evaluation of undrained rigidity index of clays ($I_R = G/s_u$) is presented, including laboratory testing, empirical correlations, and analytical methodologies. Using a hybrid spherical cavity expansion—critical state framework, an expression is derived for obtaining the operational rigidity index (I_R) directly from post-processing of CPTu data, specifically using the cone tip resistance and porewater pressure readings, or their normalized quantities. The evaluated rigidity indices are in reasonable agreement with reference laboratory tests and seismic-based in-situ approaches. The obtained values of I_R are used to calculate the yield stress (σ'_p) profiles using three separate expressions obtained from the SCE-CSSM framework, based on: (a) net cone resistance: $q_{net} = q_t - \sigma_{vo}$; (b) excess porewater pressure: $\Delta u = u_2 - u_o$; and effective cone resistance: $q_E = q_t - u_2$. The acquired value of I_R is also input into the cone bearing factor (N_{kt}) to obtain the undrained shear strength, where $s_u = q_{net}/N_{kt}$. Case studies are presented showing that the CPTu profiles of σ'_p and s_u generally agree with laboratory testing by one-dimensional consolidation and triaxial compression mode, respectively.

On the use of CPT for the geotechnical characterization of a normally consolidated alluvial clay in Hull, UK

L. Allievi, T. Curran, R. Deakin & C. Tyldsley
Arup, UK

ABSTRACT: The A63 Castle Street Improvement scheme includes an excavation approximately seven metres deep within a layer of normally consolidated alluvial clay. The collection of good quality undisturbed samples of this soil for laboratory testing proved challenging, therefore great reliance was put on site testing, especially on cone penetration tests. The procedure to calibrate and combine the data gathered from the field CPT and dissipation tests with all the ground investigation data is described. The use of CPTs provided a continuous profile with depth of the geotechnical properties of this extremely variable layer. Additional considerations are given on the effects of the use of a u1 or u2 piezocone position for the measurement of pore water pressure.

New vibratory cone penetration device for in-situ measurement of cyclic softening

D. Al-Sammarraie, S. Kreiter, F.T. Stähler, M. Goodarzi & T. Mörz
MARUM-Center for Marine Environmental Sciences, University of Bremen, Bremen, Germany

ABSTRACT: A new type of vibratory cone penetration device has been developed to improve the geotechnical in-situ methods for evaluating the dynamic and cyclic properties of the soil. The new device controls displacement amplitudes in real time and is designed to test various cyclic loads that cover different configurations of vibratory pile installation and different magnitudes of earthquakes. Some soil layers reacted strongly to the induced cyclic loads with high reduction ratios *RR* of the cone resistance, while other layers did not react to cyclic loading. It was also found that even a distance of 50 cm does not guarantee a good correlation between CPT and Vibro-CPTu. Therefore two or more pairs of CPT and Vibro-CPTu should be conducted at close distances in order to minimize misinterpretations of the data, which are caused by small scale geological structures, such as the presence of inclined layers, cross bedding, or other heterogeneities.

Estimation of geotechnical parameters by CPTu and DMT data: A case study in Emilia Romagna (Italy)

S. Amoroso
Istituto Nazionale di Geofisica e Vulcanologia, L'Aquila, Italy

P. Monaco
University of L'Aquila, L'Aquila, Italy

L. Minarelli
Geotema srl, University of Ferrara spin-off company, Ferrara, Italy

M. Stefani
University of Ferrara, Ferrara, Italy

D. Marchetti
Studio Prof. Marchetti, Rome, Italy

ABSTRACT: The aim of the paper is to illustrate and compare the results of CPTu and DMT tests carried out for a case study at Mirabello (Ferrara Province, Italy), a small municipality strongly affected by liquefaction phenomena during the 2012 Emilia earthquakes. The research site, that did not liquefy, is composed by silts and sandy silts in the shallow subsurface and by clays in lower layers. The paper compares the geotechnical parameters estimated from CPTu and DMT tests. The study also investigates the CPTu-DMT correlations according to available references (Robertson 2009, Ouyang & Mayne 2017).

Cone penetration testing on liquefiable layers identification and liquefaction potential evaluation

E. Anamali
In Situ Site Investigation, Balkans

L. Dhimitri & D. Ward
In Situ Site Investigation, UK

J.J.M. Powell
Geolabs Limited, UK

ABSTRACT: This paper presents the results of liquefaction potential evaluation for various sandy sites, with high seismic risk, where soil liquefaction is a major concern for all structures supported on these kinds of soils. There are many methods available for these calculations, based on different site investigation techniques. The Cone Penetration Test (CPTU) provides ideal data for this purpose, due to its repeatability and reliability. CPTU based methods on soil liquefaction assessment are important not only to identify liquefiable layers, but also their state in situ. This paper looks at results of liquefaction analyses on different sites susceptible to liquefaction, using CPTU based methods and also compares them with methods that are Standard Penetration Test (SPT) based, but using SPT data derived from correlations to CPTU results.

Dutch field tests validating the bearing capacity of Fundex piles

S. Van Baars & S. Rica
University of Luxembourg, Luxembourg, Grand Duchy of Luxembourg

G.A. De Nijs, G.J.J. De Nijs & H.J. Riemens
BMNED, Terneuzen, The Netherlands

ABSTRACT: In 1939, Boonstra was the first to base the tip bearing capacity of foundation piles on the CPT cone resistance. Since then, many engineers and scientists have proposed improved design methods for the bearing capacity of foundation piles, such as the Dutch Koppejan method. This method holds a mistake in the q_c-averaging method. Its tip resistance is based on the cone resistance of an assumed zone between $8D$ above the tip and $4D$ below the tip, while several researchers show that this should be in a zone between $2D$ above the tip and $8D$ below the tip. In the Netherlands, Belgium and France a field test has been performed indicating that the currently used design method in the Netherlands (the Koppejan Method) was about 30% too high for the tip resistance. This, and also the current q_c-averaging method, conflict with the findings of Boonstra, Plantema, Huizinga and White & Bolton. Besides, in the field test, the residual stresses in the piles after installation were completely ignored, in fact, not even measured. Nevertheless, the Dutch Norm Commission Geo-Engineering decided to reduce the bearing capacity of foundation piles in the Netherlands, unless other field tests prove otherwise. Since this reduction is very drastic and since no serious problems due to the use of the unreduced bearing capacity were recorded, the geotechnical contracting company Funderingstechnieken Verstraeten BV has performed field tests on six Fundex piles, and asked the engineering company BMNED to assist with these tests and the design. The aim was to prove that, at least for Fundex piles, a reduction of 30% is too much. The Fundex Pile Tests in Terneuzen show that, especially for the grouted Fundex piles, the pile type factor should not be reduced in combination with the current q_c-averaging method.

Estimation of spatial variability properties of mine waste dump using CPTu results—case study

I. Bagińska & M. Kawa
Wrocław University of Science and Technology, Poland

W. Janecki
Geosoft sp. z o.o., Poland

ABSTRACT: The paper deals with application of CPTu test results for the probabilistic modeling of lignite mine waste dump soils. The statistical measures are obtained using the results from four CPTu tests performed in a close range in the lignite mine dumping ground in Bełchatów (Central Poland). Both the tip resistance q_c as well as local friction f_s are tested. Based on the mean values, standard deviations and new classification nomogram of measured quantities, the specific zones in the dumping site profile are distinguished. For all three zones, based on normalized de-trended values for q_c and f_s both vertical and horizontal scales of fluctuation are estimated. The obtained results allow the description of dumped soil parameters using Random Fields.

Strength parameters of deltaic soils determined with CPTU, DMT and FVT

L. Bałachowski, K. Międlarz & J. Konkol
Faculty of Civil and Environmental Engineering, Gdańsk University of Technology, Gdańsk, Poland

ABSTRACT: This paper presents the results of soil investigation in soft, normally consolidated organic soil in the estuary of Vistula river. The analysis concerns clayey mud and peat layers interbedded with loose to medium-dense sands. Several Cone Penetration Tests with pore water measurement (CPTU), Dilatometer Tests (DMT) and Field Vane Tests (FVT) were performed on the testing site. The cone factor N_{kt} was estimated using the results of FVT and peak values of undrained shear strength. The proposition of N_{kt} change due to normalized Friction ratio (F_r) has been given. The possible value of cone factor $N_{\Delta u}$ for organic soils was also suggested. The undrained shear strength of Jazowa clayey mud and peat have been compared with DMT estimates proposed by various researchers. Finally, the structural soil sensitivity of Jazowa deltaic soils was determined on the basis of residual shear strength obtained from FVT and the measurements of sleeve friction.

Estimation of the static vertical subgrade reaction modulus k_s from CPT

N. Barounis & J. Philpot
Cook Costello, Christchurch, New Zealand

ABSTRACT: A methodology for the estimation of the static vertical subgrade reaction modulus (k_s) for cohesionless soils from the Cone Penetration Test (CPT) has been introduced in 2013 (Barounis et al.) and 2015 (Barounis and McMahon) and has recently been integrated (Barounis and Philpot, 2017). In this paper, the conclusions from the early two papers are utilized for developing an integrated methodology based on the correlation between q_c and N_{60} (Robertson, 2012). The fundamental concepts and the theory of the proposed methodology are presented with a step-by-step procedure in this paper. The methodology returns one value termed K_F, which is the equivalent spring stiffness for any foundation depth and shape under consideration. The methodology produces values that are as conservative as the traditional SPT approach proposed by Scott (Scott, 1981). The methodology is applied on numerous sandy sites in New Zealand for different foundation typologies.

Estimation of in-situ water content and void ratio using CPT for saturated sands

N. Barounis & J. Philpot
Cook Costello, Christchurch, New Zealand

ABSTRACT: The CPT is used extensively for site characterization, soil profiling, determination of groundwater conditions and the estimation of geotechnical parameters. The geotechnical parameters that can be estimated by using the CPT include bulk unit weight, shear strength and stiffness, among many others. Best practice suggests the CPT is to be used in combination with laboratory testing, when the budget and timeframes allow for such testing to be undertaken. For low risk projects in New Zealand, such testing is not undertaken due to a restricted budget. Typical soil classification tests, such as water content, bulk unit weight, sieve analysis and plasticity index, are not commonly performed for such projects. These parameters are important for characterizing the soil behavior, in both static and dynamic conditions. This paper proposes a methodology for estimating the in-situ water content, void ratio, dry unit weight and porosity from CPT for saturated sands.

NMO-SCTT: A unique SCPT tomographic imaging algorithm

Erick Baziw & Gerald Verbeek
Baziw Consulting Engineers Ltd., Vancouver, Canada

ABSTRACT: Seismic Cone Penetration Testing (SCPT) is an important geotechnical testing technique for site characterization that provides low strain ($< 10^{-5}$) in-situ interval compression (Vp) and shear (Vs) wave velocity estimates. Baziw Consulting Engineers has invested considerable resources in advancing the art of SCPT, and in this paper a newly developed Normal Moveout Seismic Cone Tomographic Testing (NMO-SCTT) algorithm is introduced. This algorithm allows for two dimensional imaging of the subsurface stratigraphy by processing acquired seismic trace arrival times derived with increasing source-sensor radial offsets. This dramatically increases the ability to characterize near-surface stratigraphy, which is very important for accurate liquefaction assessment. As opposed to crosshole tomography, the NMO-SCTT does not require any significant site disturbance aside from a single SCPT sounding, thereby greatly reducing the cost and the environmental impact. This paper outlines the mathematical and algorithmic details of the NMO-SCTT algorithm, which builds upon BCE's established FMDSM algorithm. As such it incorporates Fermat's principle when estimating SCPT interval velocities. In addition a real SCPT data tomographic data set is presented using SCPT seismic data that was acquired at offsets of 1.85 m, 5 m and 10 m, and down to a depth of 20.5 m.

Effect of piezocone penetration rate on the classification of Norwegian silt

A. Bihs & S. Nordal
Norwegian University of Science and Technology, Trondheim, Norway

M. Long
University College Dublin, Ireland

P. Paniagua
Norwegian Geotechnical Institute, Trondheim, Norway

A. Gylland
Multiconsult, Trondheim, Norway

ABSTRACT: Interpretation of Cone Penetration Tests (CPTU) in intermediate soils is complex due to partially drained conditions during penetration. In order to gain more insight into the material behavior of silt during a CPTU, an intensive test program was carried out in the field and several soil samples were taken and analysed. In addition a set of tests were performed in the laboratory under controlled conditions using a mini-piezocone (Fugro miniature CPTU, owned by the University of Colorado). The aim is to develop an improved interpretation basis for the CPTU in intermediate soils. The present paper gives an overview over the results obtained and shows how a change in penetration rate affects the response of the CPTU-readings. The data have been plotted into existing soil classification charts and the results from the field and laboratory investigations are compared. The Schneider et al. (2007) chart seems promising in separating the results from the different rate tests.

Quantifying the effect of wave action on seabed surface sediment strength using a portable free fall penetrometer

C. Bilici, N. Stark & A. Albatal
Virginia Polytechnic Institute and State University, Blacksburg, VA, USA

H. Wadman & J.E. McNinch
Field Research Facility, Coastal Hydraulics Laboratory, USACE, Duck, NC, USA

ABSTRACT: The surficial sediment strength of nearshore zone sands has previously been correlated to wave conditions in a qualitative manner. In this study, a coefficient of wave impact on sediment surface strength (CWS) is proposed to enable a quantitative assessment. The CWS was tested along a cross-shore transect in Duck, NC. The geotechnical input parameters for the CWS were measured using a portable free fall penetrometer. The offshore wavelength was extracted from a WaveRider buoy deployed in 17 m, and the water depth was derived from the penetrometer's pressure transducer. The results showed that the CWS coefficient reflects the change in sediment strength with wave-seabed interaction. Along the profile, the CWS ranged from 0.005 to 1.627. The CWS slightly increased at water depths shallower than the depth of closure (~4.5 m) and drastically increased at water depths shallower than breaker depth (~2 m), reaching a maximum value of 1.6.

Interpretation of seismic piezocone penetration test and advanced laboratory testing for a deep marine clay

M.D. Boone
Black and Veatch, Walnut Creek, California, USA

P.K. Robertson
Gregg Drilling & Testing, Inc. Signal Hill, California, USA

M.R. Lewis
Bechtel Corporation, Evans, Georgia, USA

D.E. Gerken
Bechtel Corporation, San Francisco, California, USA

W.P. Duffy
Bechtel Corporation, London, England, UK

ABSTRACT: A nearshore site investigation in a deep marine clay was performed in British Columbia, Canada. Investigation boreholes and seismic piezocone penetration test (SCPTu) soundings, including both seabed deployed and top push tests, were advanced to depths of greater than 135 meters below mudline. This paper presents the interpretation of the SCPTu data together with the results of laboratory testing on thin-wall (intact) tube samples from adjacent companion boreholes. Laboratory testing included index tests, consolidation tests, consolidated Direct Simple Shear (DSS), K_0 Consolidated Undrained Triaxial Compression tests (K_0CUTXC), and combined Resonant Column Torsional Shear (RCTS) tests. The results of the laboratory testing aided the interpretation of the SCPTu data including the development of site-specific correlations to parameters such as shear wave velocity and undrained shear strength.

CPT-based liquefaction assessment of CentrePort Wellington after the 2016 Kaikoura earthquake

J.D. Bray
University of California at Berkeley, Berkeley, CA, USA

M. Cubrinovski & C. de la Torre
University of Canterbury, Christchurch, New Zealand

E. Stocks
Tonkin & Taylor, Ltd., Wellington, New Zealand

T. Krall
CentrePort Ltd., Wellington, New Zealand

ABSTRACT: The Wellington CentrePort experienced liquefaction of reclaimed land and liquefaction-induced ground deformations that led to building and wharf damage in the 2016 M_w7.8 Kaikoura earthquake. There was evidence of lateral spreading in the fills behind the pile-supported wharves and liquefaction-induced settlement in the soils surrounding buildings supported on deep foundations. Previous attempts to perform CPTs at the port had limited success. In the reclaimed land constructed by end-dumping of quarried soils there are large gravel-size and boulder-size particles overlain by a 3-m⁺ thick, dense compacted earth crust covered with a thick pavement. Use of a larger, more robust cone provided reliable subsurface data that characterized well the reclaimed materials at the port. Insights garnered from this site investigation of the gravelly soils are shared.

Use of CPTu for design, monitoring and quality assurance of DC/DR ground improvement projects

Aymen Brik
Trevi Ground Engineering, UAE

Peter K. Robertson
Gregg Drilling & Testing Inc., USA

ABSTRACT: CPTu has proven to be an important tool with a multitude of uses throughout the life cycle of a ground improvement project, starting from design, execution and verification. The aim of this paper is to present several aspects of using the CPTu results in DC/DR ground improvement practice, including: exploration of pre-improvement sub-soil conditions, evaluation of the soil compactability and selection of the most suitable ground improvement solution based on CPT SBT_n-I_c ranges, anticipation of the required compaction energy according to soil type and required level of ground stiffness. Several ground improvement case histories shall be presented to illustrate these different aspects.

Keywords: Dynamic Compaction, Dynamic Replacement, CPTu, soil Compactability, compaction energy, cost effective

Cost effective ground improvement solution for large scale infrastructure project

Aymen Brik
Trevi Ground Engineering, UAE

David Tonks
Coffey, Technical Director, UK

ABSTRACT: A safe and cost-effective solution has been applied for large scale infrastructure project in Dubai (UAE), with the use of locally available sands material to create Dynamic Replacement (DR) reinforcement columns, enabling the construction of the planned development directly on the improved ground without need for deep foundations or soil replacement measures. Significant savings in terms of construction time and cost have been achieved. Cone Penetration Testing (CPT) has been used in this project as the best suited and most reliable tool for design and monitoring of DC-DR projects. It provides valuable information and serves as a basis for design, monitoring and optimization of the ground improvement works. This paper briefly describes the use of CPT for the optimization of the ground improvement solution as well as for the entire project life-cycle.

Keywords: Dynamic Replacement, carbonate sand, shell, economical, correlation, CPT, PMT

# Evaluation of existing CPTu-based correlations for the deformation properties of Finnish soft clays

Bruno Di Buò, Juha Selänpää & Tim Länsivaara
Tampere University of Technology (TUT), Tampere, Finland

Marco D'Ignazio
Norwegian Geotechnical Institute (NGI), Norway

ABSTRACT: An extensive research program for soil testing has been conducted on five soft clay deposits located in Finland. This research project aims to collect data from high quality *in-situ* and laboratory tests and derive correlations for the strength and deformation properties specific to Finnish clays. In the literature, several authors have proposed correlation models for the deformation properties of soft clays based on CPTu measurements. However, such models are often calibrated for a specific site or a specific soil type. Consequently, these models must be validated before applying them to different soil conditions. In this paper, existing correlations for the deformation properties of soft clays based on CPTu data are compared to the test results from the investigated sites. The validity of the existing models is assessed for Finnish clays by evaluating their bias and uncertainties.

Ultimate capacity of the drilled shaft from CPTu test and static load test

M.A. Camacho & J. Mejia
Independent Engineer, Cochabamba, Bolivia

C.B. Camacho, W. Heredia & L.M. Salinas
Universidad Mayor de San Simón, Cochabamba, Bolivia

ABSTRACT: The ultimate capacity for the design of a drilled shaft is affected by several components, among them: the construction methodology, materials used, field conditions of the stratigraphy and the final constructive conditions. These components directly increase the uncertainty about the interaction between the soil and the drilled shaft, since the empirical methods of calculating the ultimate capacity from field tests assume ideal characteristics, both soil and drilled shaft. Therefore, the AASHTO LRFD standard recommends carrying out a load test in order to verify the predicted capacity of the drilled shaft. This article presents the comparison of the ultimate capacity predicted (Qp) by methods from the CPTu test with that capacity measured (Qm) through a static load test in a drilled shaft built in Cochabamba, Bolivia.

Interpreting properties of glacial till from CPT and its accuracy in determining soil behaviour type when applying it to pile driveability assessments

A. Cardoso & S. Raymackers
GeoSea-DEME Group, Zwijndrecht, Belgium

J. Davidson
Cathie Associates, Newcastle, UK

S. Meissl
Orsted Energy Wind Power, Gentofte, Denmark

ABSTRACT: Glacial till comprises a mixture of material which has been transported by a glacier, such as clay, silt, sand, gravel, cobbles and even boulders. Due to this widespread and variable material, accurate driveability predictions can be very difficult to begin with. This study presents a driveability assessment that was performed for the installation of monopiles at a site located offshore of England's East coast where the Swarte Bank glacial till formation can be found. Pile driveability back-calculations showed the influence of assessing the soil behaviour type, the accuracy of interpreting properties of the glacial till from CPTs when applying it to pile driveability assessment, and how CPTs should be treated with more value than borehole description when assessing soil resistance to driving on Swarte Bank glacial till.

Variable rate of penetration and dissipation test results in a natural silty soil

R. Carroll & P. Paniagua
Norwegian Geotechnical Institute, Trondheim & Oslo, Norway

ABSTRACT: Variable rate of penetration over 1.2 to 1.5 m intervals were carried out in a natural clayey silt followed by dissipation tests. The tests are grouped into two main sets: an upper set from 5 to 6.5 m and a deeper set from 8.5 to 10.2 m. Index, strength and consolidation parameters are presented for reference to soil behavior and classification. This paper investigates the effect of rate on u_2, q_t and B_q using penetration rates of 2, 20, 100 and 320 mm/s. Similarly the effect of rate was investigated for assessment of dissipation tests and estimation of the time for 50% dissipation (t_{50}). Dissipation tests were predominantly dilatory at all rates and depths. Drainage conditions were evaluated at the different rates using the normalised rate of penetration (V) calculated using one method to estimate the horizontal coefficient of consolidation (c_h). Reference to B_q as a guide of drainage conditions is discussed together with V. A total of six methods to estimate t_{50} and subsequently c_h were used in this study, calculated t_{50} values are presented for all methods. One method is used for presentation of trends and consideration of which methods may yield the most representative c_h values is discussed in relation to laboratory c_v.

Rapid penetration of piezocones in sand

S.H. Chow, B. Bienen & M.F. Randolph
Centre for Offshore Foundation Systems, The University of Western Australia, Australia

ABSTRACT: This study investigates the change in penetration resistance across different drainage regimes in sand using centrifuge piezocone tests. The model piezocone was jacked at various penetration rates into saturated loose and dense silica sand at gravitational acceleration of 50 g. In order to achieve a wider drainage regime, the sand was saturated using both water and a viscous pore fluid (methocel cellulose ether with viscosity 715 times higher than water). The results indicate the net cone resistance increases with increasing non-dimensional velocity in dense sand, but reduces with increasing non-dimensional velocity in loose sand. This rate dependency in sand can be captured using a simple harmonic backbone curve.

Applying Bayesian updating to CPT data analysis

S. Collico, N. Perez & M. Devincenzi
Igeotest, Figueres, Spain

M. Arroyo
Department of Geotechnical Engineering, UPC, Barcelona, Spain

ABSTRACT: Evaluation of geotechnical parameters on a project site is a necessary step in geotechnical engineering. However, due to the inherent variability of soil properties and the lack of data, many unavoidable uncertainties arise during a site-specific geotechnical characterization. This challenging task can be addressed under the Bayesian framework. The aim of this paper is to apply the Bayesian approach to a reference example of friction angle evaluation in sand, using the Bayesian Equivalent Sample Toolkit (BEST). BEST is an Excel VBA program for probabilistic characterization of geotechnical properties. In particular, in this study the statistical analysis has been performed using CPT tests from reference field studies. The results obtained for one case study involving CPT are discussed.

Geotechnical characterization of a very soft clay deposit in a stretch of road works

R.Q. Coutinho & H.T. Barbosa
Federal University of Pernambuco – UFPE, Recife, Brazil

A.D. Gusmão
Polytechnic School of the University of Pernambuco – UPE, Recife, Brazil

ABSTRACT: The practice of piezocone testing for geotechnical investigations is advantageous for rapidly obtaining parameters, continuous assessment of a soil profile and ability to estimate various geotechnical parameters. However, its results require attention regarding its implementation and interpretation, especially when the aim is to perform a geotechnical characterization of an area or deposit. This paper addresses in situ tests (SPT, vane and piezocone tests) and laboratory tests of a stretch of road works located in Pernambuco, Brazil. Stratigraphic classification, compressibility and strength parameters were obtained through piezocone tests and compared to laboratory (oedometer and triaxial) and in situ (SPT, vane) tests as benchmarks. The final results were discussed within the context of results in the literature, including the results of the Recife (Coutinho 2007) and Suape (Coutinho and Bello 2012) soft clays, confirming the potential of the piezocone test to obtain good prediction of geotechnical parameters in these soft clay deposits with correlations suited to the local/regional experience.

Analysis of drainage conditions for intermediate soils from the piezocone tests

R.Q. Coutinho & F.C. Mellia
Federal University of Pernambuco – UFPE, Recife, Brazil

ABSTRACT: This paper presents an analysis of the drainage conditions in intermediate soils, as they can be loaded by piezocone tests, at a standard rate of 20 mm/s, and may present a partially drained condition. This situation requires caution due to the lack of theoretical methodology for the interpretation of data resulting from the test. In this context, the objective of this work was to evaluate the drainage conditions in compacted intermediate soils coming from the "Barreiras Formation" geological unit present in a landfill located in the Brazilian municipality of Itapissuma, Pernambuco state (PE). Two campaigns of piezocone were carried out, with the first campaign consisting of three tests at a standard rate of 20 mm/s (CPTu-01, CPTu-02 and CPTu-03) and the second campaign consisting of two tests at a penetration rate of 10 mm/s (CPTu-01 A and CPTu-03 A) and one test at the rate of 30 mm/s (CPTu-01B). In addition, characterization and oedometer tests on undisturbed samples under flooded and natural moisture conditions near the sites of the piezocone tests contributed to the determination of the coefficient of consolidation. From the reported results, it was concluded that the studied soil, when loaded at the different cone penetration rates, presented undrained behavior.

Behaviour of granitic residual soils assessed by SCPTu and other in-situ tests

N. Cruz & J. Cruz
Mota-Engil—Engenharia e Construção, S.A., Porto, Portugal

C. Rodrigues
Department of Civil Engineering, Polytechnic Institute of Guarda, Portugal

M. Cruz
LEMA, Mathematical Engineering Lab, ISEP, School of Engineering, Polytechnic of Porto, Portugal

S. Amoroso
Istituto Nazionale di Geofisica e Vulcanologia de Italia, Rome, Italy

ABSTRACT: Residual soils resulting from weathering processes cannot be well modelled by the classical theories of Soils Mechanics, creating several difficulties in the interpretation of in-situ test results and in the consequent geotechnical design. In fact, the presence of a cementation structure generates an extra strength component materialized by the presence of a cohesive intercept in the Mohr-Coulomb failure envelope. Moreover, when suction is present the interpretation of in-situ data becomes even more complicated, since it increases this cohesive intercept mixing both (cementation structure and suction) in one final result. In granular soils, as it is the case of granitic residual soils, three components of strength have to be considered, namely the shear resistance angle, cohesion and suction, which only can be solved if more than one field or laboratory measurement is obtained. As a consequence, Piezocone (CPTu), Marchetti Dilatometer (DMT) and Pressuremeter (PMT) tests may be considered appropriate to characterize this type of soils, while dynamic tests such as Standard Penetration (SPT) or Dynamic Probing Super-Heavy (DPSH) tests cannot be effective in this determination. On the other hand, cementation structure also affects deeply the stiffness behaviour, deviating from typical response of transported soils represented by Classical Soil Mechanics.

In the last two decades a big effort has been made in Portugal to study and characterize this kind of residual massifs in granitic environments (quite common in the North and Centre of Portugal), namely through DMT and CPTu data, materialized by several research frameworks and its consequent publications. In this paper, a methodology to evaluate the strength of these residual soils by means of SCPTu tests is presented and discussed, leading to a set of new correlations that allows for the evaluation of both cohesive and frictional contributions, hardly possible with the available interpretation models. The work was developed on the Polytechnic Institute of Guarda (IPG) experimental site, which was also previously used in the calibration of SDMT tests for the characterization of these granitic soils.

Rate effect of piezocone testing in two soft clays

F.A.B. Danziger, G.M.F. Jannuzzi & A.V.S. Pinheiro
Federal University of Rio de Janeiro, Brazil

M.E.S. Andrade
Technological Federal University of Paraná, Brazil

T. Lunne
Norwegian Geotechnical Institute, Norway

ABSTRACT: The Onsøy and Sarapuí II soft clays have been studied for a number of years. In particular, piezocone tests with different rates have been conducted at both sites. Since the plasticity index and the coefficient of consolidation—which play important roles in the rate effect—are different in Onsøy and Sarapuí II clays, an interesting comparison is possible. A bowl shaped curve was obtained for the normalized cone resistance versus penetration rate in the case of Sarapuí II clay. The cone resistance versus rate curve can be explained by the pore pressure trends, especially from the u_1 trend. Since only three rates have been used for Onsøy clay, a more complete picture cannot be obtained. However, the use of the normalized rate (or velocity), even using c_h from piezocone dissipation data, was not capable to unify the resistance-rate data for the two clays tested. Adaptations in regular rigs, allowing much smaller penetration rates, and small diameter cones, are necessary if drained conditions are to be achieved when conducting piezocone tests in deposits like Sarapuí II or Onsøy clays. No rate effect was found for penetration rates greater than the standard rate, unlike expected, due to the high I_p values in both deposits. Tests with higher rates are necessary to properly evaluate the role of I_p on rate effect on cone resistance.

A modified CPT based installation torque prediction for large screw piles in sand

C. Davidson, T. Al-Baghdadi, M. Brown, A. Brennan & J. Knappett
University of Dundee, Dundee, Scotland

C. Augarde, W. Coombs & L. Wang
Durham University, Durham, UK

D. Richards & A. Blake
University of Southampton, Southampton, UK

J. Ball
Roger Bullivant Limited, UK

ABSTRACT: Screw piles have been suggested as an alternative foundation solution to straight-shafted piles for jacket supported offshore wind turbines in deep water. The significant environmental loads in the marine environment will require substantially larger screw piles than those currently employed in onshore applications. This raises questions over the suitability of current design methods for capacity and installation torque. This paper aims to address this issue by presenting a screw pile installation torque prediction method based on cone resistance values from Cone Penetration Test (CPT) data. The proposed method, developed using centrifuge modelling techniques in dry sand, provides accurate predictions of installation torque for both centrifuge and field scale screw piles. Furthermore, unlike existing CPT-torque correlations, the proposed method is shown to be applicable to multi-helix screw piles.

Effects of clay fraction and roughness on tension capacity of displacement piles

L.V. Doan & B.M. Lehane
School of Civil, Environmental and Mining Engineering, The University of Western Australia, Australia

ABSTRACT: This paper examines the effect of clay fraction and roughness on the relationship between the ratio of the CPT end resistance (q_c) to unit shaft friction (τ_f) for displacement piles in sand. Tension load tests were performed on smooth and rough model piles that were jacked into sand, sand-clay mixtures and clay. Parallel constant normal load and constant normal stiffness direct shear interface tests were conducted to assist interpretation and allow separation of effects of interface friction angle and dilation on the capacities. It is shown that even a small amount of clay within a sand mass can have a major impact on the available shaft friction. The effects of pile roughness are important for developing the pile shaft capacities in these soils.

Shaft resistance of non-displacement piles in normally consolidated clay

L.V. Doan & B.M. Lehane
School of Civil, Environmental and Mining Engineering, The University of Western Australia, Australia

ABSTRACT: This paper presents results from a series of tension pile load tests on buried piles in normally consolidated kaolin and compares trends observed with shaft friction inferred from Constant Normal Load (CNL) and Constant Normal Stiffness (CNS) direct shear tests. It is shown that a CNL test with allowance for a 10% reduction in lateral effective stress during shearing provides a simple way of obtaining an accurate estimate of the available pile shaft friction. The capacities predicted using four separate of empirical methods, including more recent CPT methods, are under-predicted by a factor of between 1.5 and 2.2.

Effects of partial drainage on the assessment of the soil behaviour type using the CPT

L.V. Doan & B.M. Lehane
School of Civil, Environmental and Mining Engineering, The University of Western Australia, Australia

ABSTRACT: This paper presents the results from a series of CPTs performed at various penetration rates in kaolin, two kaolin-sand mixtures and fine silica sand. The CPTs were conducted in laboratory pressure chambers at penetration velocities varying from 0.0002 to 3 mm/s. The measured dependence of cone end resistance (q_t), sleeve friction (f_s) and pore pressure (u_2) on penetration rate is used to examine implications for the assessment of soil behavior type from CPT parameters. Element tests using triaxial and direct interfaces tests are used to assist interpretation of the test results.

Analysis of CPTU data for the geotechnical characterization of intermediate sediments

M.F. García Martínez, L. Tonni & G. Gottardi
Department DICAM, University of Bologna, Italy

I. Rocchi
Department of Civil Engineering, Danish Technical University, Denmark

ABSTRACT: The intermediate soil (e.g. silt, sandy silt, clayey silt) response at the standard cone penetration (CPT) velocity of 20 mm/s is generally partially drained, falling between that of sand and clay. As a result, a proper interpretation of CPT (or CPTU) in such mixed soils is not always straightforward. In order to properly analyse the in situ soil response and avoid incorrect estimates of soil parameters, the preliminary assessment of drainage conditions is essential. In this paper, changes in normalized CPTU measurements caused by changes in cone velocity are analysed. Penetration rate effects are assessed by means of No. 8 piezocone tests, with penetration rates ranging from about 0.9 to 61.7 mm/s. Tests were performed at a site located at the southern margin of the Po river valley (Northern Italy), where the subsoil mainly consists in a clayey silt deposit. Limitations on the applicability of some widely used empirical correlations, proposed for sands, are investigated and some preliminary results are shown.

Detection of soil variability using CPTs

T. de Gast, P.J. Vardon & M.A. Hicks
Section of Geo-Engineering, Faculty of Civil Engineering and Geosciences, Delft University of Technology, Delft, The Netherlands

ABSTRACT: The variability of soil is well known to affect the geotechnical performance of structures. As probabilistic design methods become more commonly used, the ability to measure the variability of soil becomes more important. However, by using only the point statistics of soil parameters in design (e.g. the mean and standard deviation), typically an over-estimation of failure probabilities occurs, leading to over-conservative designs. By looking at the spatial correlation (e.g. scales of fluctuation) a more accurate representation can be achieved. This paper presents a method to use vertical Cone Penetration Tests (CPTs) to detect both the vertical and horizontal scales of fluctuation. An extensive numerical and experimental investigation has been undertaken to understand how spatial variation can be estimated and to quantify the accuracy in that estimation. The impact of being able to quantify the uncertainty is illustrated via a simple slope stability example.

MPM simulation of CPT and model calibration by inverse analysis

P. Ghasemi & M. Calvello
Department of Civil Engineering, University of Salerno, Fisciano, Italy

M. Martinelli & V. Galavi
Deltares, Delft, The Netherlands

S. Cuomo
Department of Civil Engineering, University of Salerno, Fisciano, Italy

ABSTRACT: The Material Point Method (MPM) has been employed in the study of many geotechnical large deformation problems. This study aims at showing: the effectiveness of MPM in simulating a CPT, and the possibility to calibrate the model input parameters by inverse analysis. The MPM schematization adopted to implement the CPT boundary value problem uses a moving mesh concept to model the cone as a rigid body penetrating into the soil. A gradient-based non-linear regression analysis is used to calibrate the input parameters of the constitutive law adopted to model the soil. To this aim, a synthetic case study has been set up, using the model results of a "base" simulation as observations and, in particular, the cone resistance values computed at 24 different depths. The results of the performed sensitivity, parametric and inverse analyses highlights few important aspects related to the use of optimization algorithms to calibrate the input parameters of MPM models of CPTs.

Challenges in marine seismic cone penetration testing

P. Gibbs, R.B. Pedersen & L. Krogh
Ørsted Wind Power, Copenhagen, Denmark

N. Christopher & B. Sampurno
Fugro, Wallingford, UK

S.W. Nielsen
COWI, Copenhagen, Denmark

ABSTRACT: Current advancements in foundation design methodologies have given a new focus on acquiring *in situ* G_0 data to compliment current laboratory testing methods and to provide G_0 profiles for design in a parameter bound framework. Seismic CPTU (SCPTU) data was acquired from sites within the North Sea both in drilling and non-drilling mode. An assessment of the data showed significant scatter in the interpreted results. Current standards provide little guidance on what is 'accurate and reliable' data, hence an assessment was initiated to better understand the reasons for the scatter observed and what is to be considered reliable data. This paper discusses the assessments that have been undertaken and covers: the seismic testing set-ups, how the data was acquired; how the data was interpreted; offers suggestions on what can be causing the scatter; and provides commentary on how to gain confidence in the data acquired. This study highlights the need for greater understanding of SCPTU data and closer client-contractor collaboration. It is recommended that project specifications clearly define SCPTU reporting beyond the standard requirements, where comprehensive information of the seismic data acquired and the methods used to derive the v_s is needed to improve confidence on data evaluation and interpretation for the design purpose.

Keywords: seismic velocity, CPT, small strain shear modulus

Numerical simulation of cone penetration test in a small-volume calibration chamber: The effect of boundary conditions

M. Goodarzi
MARUM—Center for Marine Environmental Sciences, University of Bremen, Bremen, Germany
Geo-Engineering.org GmbH, Bremen, Germany

F.T. Stähler & S. Kreiter
MARUM—Center for Marine Environmental Sciences, University of Bremen, Bremen, Germany

M. Rouainia
School of Engineering, Newcastle University, Newcastle upon Tyne, UK

M.O. Kluger
MARUM—Center for Marine Environmental Sciences, University of Bremen, Bremen, Germany

T. Mörz
MARUM—Center for Marine Environmental Sciences, University of Bremen, Bremen, Germany
Geo-Engineering.org GmbH, Bremen, Germany

ABSTRACT: In this study, laboratory Cone Penetration Tests (CPTs) were conducted in the MARUM Calibration Chamber (MARCC) with three lateral boundary conditions: (BC), constant stress, constant strain and the simulated field conditions with constant stiffness. Cuxhaven-Sand was studied in the chamber tests and tip resistance-relative density ($q_c - D_r$) relationships were generated for each BC. Laboratory experiments were carried out to estimate the mechanical properties of the Cuxhaven-Sand. Multiple numerical analysis have then been undertaken to simulate the calibration chamber results. First, the soil model was calibrated against laboratory soil parameters and a CPT result of the calibration chamber with fixed lateral boundaries, then, a numerical penetration analysis in an infinite soil mass was performed to evaluate the implemented constant stiffness boundary condition in the chamber. Good agreement between experimental and numerical cone resistances demonstrates the possibility of using the advanced small volume MARCC for producing controlled CPT results applicable in true field test conditions.

Transition- and thin layer corrections for CPT based liquefaction analysis

J. de Greef & H.J. Lengkeek
Witteveen+Bos, Deventer, The Netherlands

ABSTRACT: Liquefaction potential of the subsoil is often assessed by performing an automated CPT based liquefaction triggering procedure. Due to a bias in the measured cone penetration resistance erroneous prediction of the liquefaction potential can be expected at the transition of soft, non-liquefiable and stiffer, liquefiable layers. This too is the case for relatively thin liquefiable layers interbedded in soft soil layers. A simple procedure is proposed that provides a solution to both issues in an automated liquefaction potential screening.

Soil classification of NGTS sand site (Øysand, Norway) based on CPTU, DMT and laboratory results

A.S. Gundersen, S. Quinteros, J.S. L'Heureux & T. Lunne
Norwegian Geotechnical Institute (NGI), Oslo, Norway

ABSTRACT: The Norwegian GeoTest Site project (NGTS) established five research sites in Norway in 2016. The sites are referred to as sand, soft clay, quick clay, silt and permafrost. The project is funded by the Research Council of Norway and the aim of the project is to establish, characterize, share digital data and manage the use of the test sites in the coming 20 years. The sites are open to other researches for developing and calibrating new tools and techniques. The focus of this paper is the soil classification of the NGTS sand site at Øysand based on Cone Penetration Tests (CPTUs) and Dilatometer Tests (DMT). The fluvial and deltaic deposit at Øysand consists of a 20–25 m fine silty sand with occasionally high content of gravel. The deposit is generally normally consolidated in loose to medium dense states. The *in situ* test data is further supported by laboratory test results from a 20 m long and continuous borehole. This paper presents the results of two CPTUs and one DMT in addition to laboratory test results, all from the same location at the research site. The prediction of *soil behavior type* and unit weights from CPTU and DMT tests, based on existing correlations, are compared qualitatively to the soil classification from grain size distribution and unit weights from laboratory measurements.

Numerical study of anisotropic permeability effects on undrained CPTu penetration

L. Hauser & H.F. Schweiger
Institute of Soil Mechanics, Foundation Engineering and Computational Geotechnics, Graz University of Technology, Graz, Austria

L. Monforte & M. Arroyo
Department of Geotechnical Engineering, Universitat Politècnica de Catalunya – BarcelonaTech, Barcelona, Spain

ABSTRACT: The numerical simulation of Cone Penetration Tests (CPT) is a challenging field in geotechnics. Both the underlying physical model and the numerical method need to be capable of taking into account large deformations and displacements within the problem domain. Researchers at the Polytechnic University of Catalonia (UPC) and the International Center for Numerical Methods in Engineering (CIMNE) have developed a Particle Finite Element Method (PFEM) code named G-PFEM which conducts fully coupled analysis of penetration problems in saturated porous media. An updated Lagrangian description is used in order to formulate the governing equations. The PFEM is based on frequent remeshing of critical regions of the problem domain adding additional computational cost to the solving process. Therefore, the use of linear elements in combination with a stabilized mixed formulation of the governing equations helps to reduce the computational effort and at the same time cope with the phenomenon of volumetric locking associated with linear elements. Within the present work, anisotropic permeability is introduced and tested by means of consolidation of an elastic soil layer. Furthermore, recalculations of an available in-situ CPT are performed allowing the examination of the influence of changing boundary conditions, such as anisotropic permeability of the soil or penetration velocity of the cone, on the measured tip resistance, sleeve friction and pore water pressure. It was found that the recalculation with G-PFEM provides comparable results for undrained conditions and anisotropic permeability.

Evaluating undrained shear strength for peat in Hokkaido from CPT

H. Hayashi & T. Yamanashi
Civil Engineering Research Institute for Cold Region (CERI), Sapporo, Japan

ABSTRACT: In case of evaluating stability of soft ground using circular slip analysis, it is very important to determine the undrained shear strength (S_u) of the ground. Meanwhile, fibrous and high organic peat is distributed widely in Hokkaido, the northernmost island of Japan. As peat is accumulated heterogeneously, unconfined compression tests, vane shear tests and other tests performed for only a few samples lack validity. The electric Cone Penetration Test (CPT) is more reasonable and valid than these, in that CPT can estimate average S_u from in-situ tests, which continuously provide information. A series of K_0 consolidated-undrained triaxial compression tests (K_0CUC tests) on undisturbed peat soil collected at several sites in Hokkaido was conducted. Also the CPT was performed at the same sites. This report describes the relationship between S_u obtained from the K_0CUC tests and CPT cone resistance (q_t), and proposes a method for estimating S_u in peat soil from q_t.

Interpreting improved geotechnical properties from RCPTUs in KCl-treated quick clays

T.E. Helle
Directorate of Public Roads, Trondheim, Norway

M. Long
University College Dublin, Dublin, Ireland

S. Nordal
Norwegian University of Science and Technology, Trondheim, Norway

ABSTRACT: In January 2013, Potassium Chloride (KCl) wells were installed in a highly sensitive quick clay deposit at Dragvoll, Trondheim, Norway. The geotechnical properties significantly improved in the following years due to increased salt content and changed pore-water chemistry. Several Resistivity Cone Penetration Tests (RCPTU) were conducted around the wells to map the salt-plume extent, and to interpret improved geotechnical properties. The RCPTUs effectively mapped the salt plume extent around the wells, and the increased tip resistance clearly indicated improved properties in the salt-treated clay. Commonly used correlations for Norwegian clays are used herein for interpreting the geotechnical properties in the original quick clay and the salt-treated clay. The interpreted Over Consolidation Ratio (OCR) and shear strength (c_u) are compared to laboratory determined geotechnical properties on high quality downsized Sherbrooke samples (mini-blocks). The tip resistance in the salt-treated clay increased significantly due to increased remolded shear strength and increased resistance to deformations. Improved OCR and c_u in the salt-treated clay at Dragvoll are determined from the CPTU data applying existing correlations based on bearing factor for net tip resistance (N_{kt}). The existing correlations for Norwegian clays work reasonably well, but ideally specific local correlations should be developed.

Some experiences using the piezocone in Mexico

E. Ibarra-Razo & R. Flores-Eslava
InGeum Ingeniería SA de CV, Ciudad de México, México

I. Rivera-Cruz
Thurber Engineering, Ltd., Vancouver, BC, Canada

J.L. Rangel-Núñez
Universidad Autónoma Metropolitana, Azcapotzalco, Ciudad de México, México

ABSTRACT: The use of the Cone Penetrometer Test sounding (CPT) in geotechnical exploration in Mexico goes back to the 1970s, during which time a great deal of experience and a set of empirical correlations with different mechanical properties of the soils have been obtained; however, that local experience has been limited almost exclusively to the measurement of tip resistance (qc) in soft clay deposits, especially in the Valley of Mexico. In recent years, new versions of cones have been made available in our country that allow digital measurements and visual checks on real time, not only of end resistance against depth but also of sleeve friction and pore pressure during driving, *i.e.*, the piezocono test (CPTu). In this paper, we present several case histories of the application of the digital piezocone (CPT-u) in various geotechnical environments; In soft clayey stratified soils with hard layers, in clean sands and in tailings deposits. It highlights the advantages and disadvantages of this new generation of electrical cones, mainly on aspects of field execution, equipment operation and interpretation of recorded data.

Evaluation of complex and/or short CPTu dissipation tests

E. Imre
Alternative Energy Techn, Knowledge Center, Obuda University, Budapest, Hungary

T. Schanz
Ruhr University of Bochum, Germany

L. Bates & S. Fityus
University of Newcastle, Newcastle, Australia

ABSTRACT: A case-study of evaluating nine u_2 CPTu dissipation tests (I to V types) of small t_{50} with 3 published methods is presented. One method is a one-point fitting method, based on I_r and t_{50}. The two newer methods implies automatic, multi-point curve fitting, with output for not only c but also for the initial condition, the parameter error and some reliability information. The results indicate that the dissipation tests are good tools to identify soil property/layer boundary information, the new methods can be used to study the partly drained penetration and to get c value using tests shorter than the t_{50} testing time.

New Russian standard CPT application for soil foundation control on permafrost

O.N. Isaev & R.F. Sharafutdinov
Gersevanov NIIOSP, Moscow, Russia

N.G. Volkov
Fugro Group, Moscow, Russia

M.A. Minkin & G.Y. Dmitriev
FUNDAMENTPROIECT, Moscow, Russia

I.B. Ryzhkov
BGAU, Ufa, Russia

ABSTRACT: In 2016 in Russia, a new Standard STO 36554501-049-2016 "CPT Application for Soil Foundation Control on Permafrost" was developed and enacted. The standard determines the basic requirements for CPT (Cone Penetration Testing) application for geotechnical control and investigation of the foundations located on permafrost soils. The standard applies to soils which are dispersive, natural and man-made, frozen/freezing and thawed/thawing. The composition and condition of these soils should allow continuous penetration of a penetrometer used for CPT. Many chapters of the standard are relevant and applicable for performing geotechnical site investigation for foundation design purposes. Particular attention is paid to the CPT probe applications with additional sensors.

Thermophysical finite element analysis of thawing of frozen soil by means of HT-CPT cone penetrometer

O.N. Isaev, R.F. Sharafutdinov & D.S. Zakatov
Gersevanov NIIOSP, Moscow, Russia

ABSTRACT: The paper presents the results of the FE-based numerical thermophysical analysis of geocryological factors (soil type, temperature and moisture) and constructive-technological factors (power, heating and cooling time) having an effect on the thawing area size around the HT-CPT cone penetrometer. Based on the results obtained, the empirical dependencies are proposed to predict the depth and diameter of thawing area around the HT-CPT cone penetrometer.

Large deformation modelling of CPT probing in soft soil—pore water pressure analysis

J. Konkol & L. Bałachowski
Faculty of Civil and Environmental Engineering, Gdańsk University of Technology, Gdańsk, Poland

ABSTRACT: This paper presents the results of finite element modelling with Updated Lagrangian formulation of the Cone Penetration Test in soft soil deposit located in Jazowa, Poland. The numerical calculations are carried out for homogenous, normally consolidated, organic soil layer. The Modified Cam Clay constitutive model for soft soil and Coulomb model for interface are used. The study compares the registered pore water pressure distributions for type-2 piezocone observed during in-situ penetration and corresponding numerical model. The numerical dissipation test is carried out and the results are confronted with in-situ registered data. The influence of orthotropic soil hydraulic conductivity on pore water pressure development at shoulder filter element during dissipation tests is examined. Finally, the distribution of pore water pressures around the piezocone obtained from numerical simulations is compared with high quality literature database.

The use of neural networks to develop CPT correlations for soils in northern Croatia

M.S. Kovacevic
Faculty of Civil Engineering, University of Zagreb, Zagreb, Croatia

K.G. Gavin & C. Reale
Faculty of Civil Engineering and Geosciences, Delft University of Technology, Delft, The Netherlands

L. Libric
Faculty of Civil Engineering, University of Zagreb, Zagreb, Croatia

ABSTRACT: The evaluation of soil parameters for design is best undertaken through comprehensive laboratory test programmes. However, due to sampling difficulty, time and cost constraints correlations between in-situ tests and physical-mechanical properties of soils are routinely applied in practice. This paper presents data collected from five sites in Northern Croatia at which Cone Penetration Tests (CPT) and comprehensive laboratory test data was available. One of the advantages of using CPT data in preference to other types of in-situ tests for establishing correlations, is the large volume of high-quality data available at each probe location allows for the application of advanced statistical approaches. In this paper, the use of neural networks in developing such correlations is demonstrated. Using a database of 216 data pairs, obtained from the five sites, a correlation between CPT q_c and soil unit weight is established. A validation exercise was performed in which the correlation was tested against data from the recent Veliki vrh landslide that occurred in the same geographical region as the database sites. In addition, by using the soil behaviour type index, Ic, normalised cone tip resistance, Qtn, and normalised sleeve friction, Fr, the results can be compared to correlations developed for soils from geotechnical diverse regions to check for consistency in the derived correlations.

CPT in thinly inter-layered soils

D.A. de Lange & J. Terwindt
Deltares, Delft, The Netherlands

T.I. van der Linden
Royal HaskoningDHV, Amersfoort, The Netherlands

ABSTRACT: The interpretation of CPT within intervals consisting of multiple sequences of thin soil layers holds large uncertainty. For multi-layer systems, it is expected that the cone resistance would be influenced by the layer thickness (relative to the cone diameter), the number of layers within the zone of influence and the characteristic cone resistances of the individual layers. The Dutch method for determination of the pile base resistance can be used to simulate the cone resistance in thinly inter-layered soils and to come up with correction factors as function of the aspects mentioned above. A test program (parametric study) is defined in order to validate the proposed modified Dutch method. The proposed simulation method fits reasonably well with the interim test results.

CPT based unit weight estimation extended to soft organic soils and peat

H.J. Lengkeek
Delft University of Technology, Delft, The Netherlands
Witteveen+Bos, Deventer, The Netherlands

J. de Greef
Witteveen+Bos, Deventer, The Netherlands

S. Joosten
Delft University of Technology, Delft, The Netherlands

ABSTRACT: A reliable estimate of the saturated soil weight from CPT analysis can be useful for various purposes. An often used relation that gives a reasonable first approximation is presented by Robertson & Cabal (2010). In The Netherlands very soft and highly organic soils are omnipresent and these types of soil are absent in the aforementioned relation. In this paper a new relation is proposed that can be used to estimate the saturated soil unit weight for a wider range of soils, from sands to highly organic soils.

Impact of sample quality on CPTU correlations in clay—example from the Rakkestad clay

J.S. L'Heureux, A.S. Gundersen, M. D'Ignazio, T. Smaavik, A. Kleven, M. Rømoen, K. Karlsrud, P. Paniagua & S. Hermann
Norwegian Geotechnical Institute (NGI), Oslo, Norway

ABSTRACT: As part of the zoning plan for the new 30 km long highway, E16, from Nybakk to Slomarka, an extensive laboratory and field testing campaign was conducted by NGI. The deposit along the highway is a normally to slightly overconsolidated clay with a water content in the range of 30–45% and a plasticity index ranging between 7–25%. 72 mm diameter piston samples at 70 localities were taken and CPTUs were carried out at over 120 locations. About 180 CRS oedometer tests and 360 triaxial tests were performed. These results generally show good to excellent sample quality. However, due to the interpreted lower values of undrained shear strength from local CPTU correlations, 9 block samples were retrieved at 3 locations and additional laboratory testing was performed. The active undrained shear strength obtained from the block samples was up to 53% higher. Based on this data set, correlations were established to optimize the engineering solutions for the new road and to address the impact of sample disturbance on geotechnical engineering parameters. The results lead to important economical saving for the project and highlight the need for local and high quality samples for CPTU correlations in soft and sensitive clays.

Use of the free fall cone penetrometer (FF-CPTU) in offshore landslide hazard assessment

J.S. L'Heureux & M. Vanneste
Norwegian Geotechnical Institute, (NGI), Trondheim and Oslo, Norway

A. Kopf
MARUM, University of Bremen, Germany

M. Long
University College Dublin (UCD), Dublin, Ireland

ABSTRACT: Free Fall penetrometer (FF-CPTU) testing can provide significant advantages over conventional CPTU investigation for shallow sub-surface offshore site investigations. Much work has been done on soil characterisation and on determination of the undrained shear strength (s_u) from FF-CPTU testing. However, little data has been published on analysis of FF-CPTU dissipation testing. Here, FF-CPTU data from two Norwegian fjords with evidence of recent landsliding are presented, and the techniques used to analyse and correct the data are described. At Hommelvika, relatively high residual excess pore pressures (15–17 kPa) were found in the vicinity of a large pockmark identified on the fjord bed from multibeam data. At Finneidfjord, the residual excess pore pressures are lower and are in agreement with long-term piezometer data from the area. Reliable estimates of the coefficient of consolidation (c_h) were also be obtained from the FF-CPTU dissipation tests.

Fibre optic cone penetrometer

P. Looijen, N. Parasie, D. Karabacak & J. Peuchen
Fugro, Nootdorp, The Netherlands

ABSTRACT: This paper presents a novel prototype cone penetrometer for piezocone penetration tests (CPTU). It is equipped with Fibre Optic (FO) sensors for measuring cone resistance (q_c) and sleeve friction (f_s). The FO sensors replace the classical, electric strain-gauges. Field results, laboratory calibration and uncertainty assessment show capability of achieving q_c and f_s values that are well within Application Class 1 of ISO 22476-1 "electrical cone and piezocone penetration test". This makes the FO cone penetrometer particularly suitable for use in very soft clays, silts and peat, i.e. Soil A and Interpretation H of ISO 22476-1. The cone penetrometer has potential to meet a significantly more stringent, Dutch Application Class 1+.

Some considerations related to the interpretation of cone penetration tests in sulphide clays in eastern Sweden

A.B. Lundberg
ELU Konsult AB, Stockholm, Sweden

E.A. Alderlieste
SPT Offshore BV, Woerden, the Netherlands

ABSTRACT: Holocene clays that contain a significant amount of sulphide are regularly discovered during site investigation programs carried out along the East coast of Sweden. Cone Penetration Tests (CPTs) are frequently executed during such investigations, and the mechanical properties of the soil is subsequently interpreted from the CPTs. The mechanical behaviour of the sulphide clays is somewhat similar to other high-plasticity Scandinavian clays, including shear-strength anisotropy, and a high level of strain-rate and temperature dependency. These factors make the interpretation of the soil properties complicated, since the testing conditions and the testing methods have a large influence on the soil behaviour. Sample disturbance and the temperature of laboratory direct simple shear tests also influence the interpretation, since these have been used as a reference for in-situ tests. The boundary conditions of the soil during both the in-situ and laboratory tests are consequently mixed in the interpretation, resulting in significant uncertainty about the correct soil properties. The design shear strength is typically chosen conservatively, but studies show a significant variation in the resulting cone factors for sulphide clays from empirical correlations. There are however some cases where a low undrained shear strength is not conservative, and in which design guidelines create some confusion. A case study of Cone Penetration tests in sulphide clay in Eastern Sweden is examined and some factors which influence the soil behaviour are discussed, including the strain-rate, temperature and anisotropic strength.

Effect of cone penetrometer type on CPTU results at a soft clay test site in Norway

T. Lunne, S. Strandvik, K. Kåsin & J.S. L'Heureux
Norwegian Geotechnical Institute, NGI, Oslo, Norway

E. Haugen, E. Uruci, A. Veldhuijzen, M. Carlson & M. Kassner
Norwegian Public Road Administration, NPRA, Oslo; Pagani, Italy
Geomil, The Netherlands
Geotech, Sweden
Adam Mickiewicz University in Poznan, Poland

ABSTRACT: Seven different cone penetrometers from 5 manufacturers have been used in a comparative testing program at the Norwegian GeoTest Site (NGTS) on soft clay in Onsøy, Norway. Tests with all cone types give very repeatable penetration pore pressure, u_2. When comparing tests with different cone types, six of the cones give very similar u_2 values. One cone type give consistently higher u_2 values. Measured cone resistance, q_c, generally varies somewhat more, both regarding tests with the same cone, and when comparing results of one cone type with another. Some of the cone types give good repeatability for sleeve friction, f_s, readings, while some show relatively large variation. When comparing f_s from different cone types the variation is quite large, which is in line with previous experience. An attempt has been made to understand the reasons for the large f_s variations.

Evaluation of CPTU N_{kt} cone factor for undrained strength of clays

P.W. Mayne
Georgia Institute of Technology, Atlanta, Georgia, USA

J. Peuchen
Fugro, The Netherlands

ABSTRACT: The evaluation of undrained shear strength of clays (s_u) is most often sought using the net cone resistance ($q_{net} = q_t - \sigma_{vo}$) and a cone factor ($N_{kt}$) such that $s_u = q_{net}/N_{kt}$. While site-specific calibration of N_{kt} with laboratory reference values (i.e. triaxial compression, simple shear) or field benchmark (i.e. vane) is the best approach, this requires considerable extra time and funding to accomplish. In the approach covered herein, a database involving 407 high-quality triaxial compression tests (CAUC) was used to review strengths from a wide variety of clays ranging from intact soft to firm to stiff to hard and fissured geomaterials. The study considered a total 62 clays, categorized into five groups: soft offshore, soft-firm onshore, sensitive, overconsolidated, and fissured clays. The backfigured N_{kt} factors ranged from 8 to 25 and found to decrease with pore pressure ratio, $B_q = (u_2 - u_0)/q_{net}$.

Cone Penetration Testing 2018 – Hicks, Pisanò & Peuchen (Eds)
© 2018 Delft University of Technology, The Netherlands, ISBN 978-1-138-58449-5

Applying breakage mechanics theory to estimate bearing capacity from CPT in polar snow

A.B. McCallum
Faculty of Science, Health, Education and Engineering (FOSHEE), University of the Sunshine Coast, Australia

ABSTRACT: Demand for polar infrastructure is increasing and accurate determination of bearing capacity from CPT data is increasingly useful. Breakage mechanics theory has been applied to enhance pile end-bearing capacity predictions in crushable soils. Because polar snow can also be considered a crushable geomaterial, this technique may be useful in enhancing bearing capacity predictions using polar snow CPT data. This technique was preliminarily examined to investigate its application in determining bearing capacity from polar snow CPT data. Application of contemporary soil behavioural theories to polar snow undergoing penetration does not routinely occur and insights not typically examined by glaciologists can be obtained. However, extensive controlled laboratory testing is necessary to further refine the application of this technique to accurately deduce bearing capacity from CPT in polar snow.

Empirical correlations to improve the use of mechanical CPT in the liquefaction potential evaluation and soil profile reconstruction

C. Meisina
Department of Earth and Environmental Sciences, University of Pavia, Italy

S. Stacul & D.C. Lo Presti
Department of Civil & Industrial Engineering, University of Pisa, Italy

ABSTRACT: CPT-based simplified methods are the common used approaches to determine the liquefaction hazard and they require cone penetration test with electrical tip. However, in some countries, as Italy, penetrometric tests are carried out with mechanical tip (CPTm). The cone—shape effects on sleeve friction (fs) have the greatest influence on soil classification in terms of SBT, underestimating the grain size of loose soils (e.g. sands) with respect to CPTu. An empirical correlation between the fs measured with CPTm and CPTu was tested. Moreover, another correlation was developed to determine a ΔIc value as function of the cone resistance in the case of silty sands and sandy silts non correctly identified by the SBT classification systems. The correlation was applied to tests carried out in the area interested by the 2012 Emilia earthquake (Italy), where liquefaction phenomena have occurred. The procedure makes possible to use huge existing database (CPTm) for liquefaction risk assessment.

Rigidity index (I_R) of soils of various origin from CPTU and SDMT tests

Z. Młynarek
University of Life Sciences in Poznan, Poland

J. Wierzbicki & K. Stefaniak
Institute of Geology, Adam Mickiewicz University in Poznan, Poland

ABSTRACT: The paper presents an analysis of the qualitative impact of soil properties of varied origin on the rigidity index variability. This coefficient is defined as the ratio between the shear modulus G and undrained shear strength su. In order to determine values of this coefficient in four genetically varied soil groups, CPTU and SDMT tests were performed. The article discusses the concepts of evaluating the G modulus based on the values measured with the SDMT test, as well as the coefficient $I_R' = G_0 \cdot s_u^{-1}$. The results have revealed that the significant influence on the I_R variation, apart from the characteristic soil properties, such as type of soils or plasticity index, is displayed by the cementation and over consolidation effects, and the anisotropy of macrostructure of certain sediments.

A state parameter-based cavity expansion analysis for interpretation of CPT data in sands

P.Q. Mo
China University of Mining and Technology, Xuzhou, China

H.S. Yu
University of Leeds, Leeds, UK

ABSTRACT: Cone Penetration Testing (CPT) serves as a useful in-situ tool for site investigation and soil characterization, while the end-bearing and shaft capacities of driven piles could be directly related to the CPT measurements. In terms of interpretation of CPT data, state parameter concept has been employed largely owing to its good indication of soil behaviour at different stress levels and densities. Drained cavity expansion solution in a unified state parameter model for clay and sand (CASM) is adopted in this paper for the applications to CPT in sands regarding to in-situ soil state. The effects of initial stress condition, friction angle and soil compressibility on the correlations between cone tip resistance, sleeve friction and state parameter are presented and discussed. The proposed method indicates the influence of initial state parameter on the evaluation of normalised penetration resistance, and the state parameter is directly related to the soil behaviour type index. The analysis contributes to the theoretical background of the framework for interpretation of CPT data.

Permeability estimates from CPTu: A numerical study

L. Monforte, M. Arroyo & A. Gens
Universitat Politècnica de Catalunya—BarcelonaTech, Barcelona, Spain

C. Parolini
Politecnico di Milano, Milan, Italy

ABSTRACT: Realistic numerical simulations of CPTu in soft soils are now becoming possible thanks to methods like the Particle Finite Element Method (PFEM). They may be used to systematically explore the traditional methods of CPTu interpretation, identify their limitations and propose alternatives. This paper discusses a series of tests simulation CPT installation and subsequent dissipation tests in soils represented by the Modified Cam Clay model. Simulation outputs obtained for different input permeabilities are examined to obtain direct estimates of permeability using different methods proposed in the literature; additionally, a method to estimate the hydraulic conductivity during the piezocone penetration is also used. These estimates are then compared with the known input permeability value to assess their reliability.

Pore pressure measurements using a portable free fall penetrometer

M.B. Mumtaz, N. Stark & S. Brizzolara
Virginia Polytechnic Institute and State University, Blacksburg, VA, USA

ABSTRACT: Pore pressure measurements during penetrometer deployments can complement measurements of strength resistance, and provide additional information about the soil. However, for high-velocity impacts, hydrodynamic forces lead to complex pressure recordings during the free fall and penetration. To investigate these effects, a Computational Fluid Dynamics model (CFD) has been developed. Based on initial results from the CFD model, a quasi-steady approach based on a velocity-dependent correlation is proposed to correct for dynamic effects and shape dependent pressure gradients acting on the FFP, and thus, also on the pore pressure readings. Finally, the results are discussed with regard to opportunities for future research.

Influence of soil characteristics on cone and ball strength factors: Case studies

T.D. Nguyen
Infrastructure Engineering Program, Vietnam Japan University, Hanoi, Vietnam
Formerly, Department of Civil Engineering, Dong-A University, Busan, South Korea

S.G. Chung
Department of Civil Engineering, Dong-A University, Busan, South Korea

ABSTRACT: This paper presents a comparative study on the influence of some typical soil characteristics on the cone and ball factors (N_{kt} and N_{ball}). Experimental data from CPTU and ball penetration test (BPT) on Busan (two sites), Ariake, and Mihara clays were taken into analysis. In contrary to the conclusions from existing FE analysis and experimental results at limited sites that N_{ball} is independent on rigidity index (I_r) and normalized stress difference (Δ), N_{ball} was found to clearly increase with the increase in I_r and to increase or decrease with Δ. This disagreement is attributed to different inherent characteristics of natural clays. Both N_{kt} and N_{ball} decrease pronouncedly with the increase in normalized shear strength (s_{uFV}/s'_{v0}), however the factors expose to have no definite correlations with sensitivity (S_{u-V}) and plasticity index (I_p) at the investigated sites.

A method for predicting the undrained shear strength from piezocone dissipation test

E. Odebrecht
State University of Santa Catarina, Joinville, Brazil

F.M.B. Mantaras
Geoforma Engenharia Ltd., Joinville, Brazil

F. Schnaid
Federal University of Rio Grande do Sul, Porto Alegre, Brazil

ABSTRACT: A method developed to correlate the measured piezocone dissipation excess pore-water pressure (Δu) to the soil undrained shear strength (s_u) has been presented by Mantaras et al. (2015). In the proposed approach, a mathematical closed form solution was derived linking the ratio of normalized maximum excess pore pressure and the normalized undrained shear strength. In the preset paper the predicted s_u values obtained from the from dissipation tests are calibrated against field vane shear strength in both normally consolidated and lightly overconsolidated soils (OCR of up to 5). Data from 5 cm² and 10 cm² piezocone cross section areas are evaluated to demonstrated that reported results are consistent and encourage the use of the method in engineering practice.

Realistic numerical simulations of cone penetration with advanced soil models

Zhandos Y. Orazalin
Singapore-MIT Alliance for Research and Technology, Singapore

Andrew J. Whittle
Massachusetts Institute of Technology, Cambridge, MA, USA

ABSTRACT: The analysis of soil penetration represents a challenging class of geotechnical problems due to large deformations, high gradients of the field variables (stresses, strains, pore pressures, etc.) around the penetrometer, the various drainage conditions and complex constitutive behavior of soils. Most prior research using large deformation Finite Element (FE) methods has been limited to simplified assumptions on drainage conditions and constitutive behavior. Following earlier work by Hu & Randolph (1998), we propose a finite element analysis procedure using automated remeshing and solution mapping within a commercial FE solver (ABAQUS Standard) in order to simulate quasi-static piezocone penetration using advanced effective stress soil models. Predictions of piezocone penetration using the proposed FE analyses are evaluated through comparison with undrained *steady state* analytical solutions obtained from the Strain Path Method and with field measurements from Boston. Predictions of partially drained penetration are compared with recently published 2-phase MPM analyses and with data from laboratory (1g and centrifuge) model tests in kaolin.

Calibrating NTH method for ϕ' in clayey soils using centrifuge CPTu

Z. Ouyang & P.W. Mayne
Georgia Institute of Technology, Atlanta, USA

ABSTRACT: An existing effective stress limit plasticity solution for piezocone penetration tests (CPTu) developed at the Norwegian Institute of Technology (NTH) is calibrated to evaluate the effective stress friction angle ϕ' for clayey soils subjected to centrifuge model testing. Results from previously conducted in-flight mini-piezocone tests by various institutes were compiled for study, including readings taken from 13 series of centrifuge chamber tests on artificially prepared clays, mainly kaolin and kaolinitic-silica mixtures. Effective stress friction angles from companion triaxial compression tests (CAUC or CIUC) and/or Direct Simple Shear (DSS) tests on these soil deposits were adopted as the benchmark reference to verify the reasonableness of the NTH method in assessing ϕ' at constant penetration rates, as well as a few special tests conducted at variable rates.

CPT interpretation and correlations to SPT for near-shore marine Mediterranean soils

S. Papamichael & C. Vrettos
Technical University of Kaiserslautern, Germany

ABSTRACT: CPT and SPT field data from the Mediterranean Region comprising onshore and offshore sands, silts, clays and organic soils were analyzed and cross-correlated for the purpose of this research. Laboratory tests on retrieved samples were utilized to aid the evaluation of the applicability of CPT as a site investigation tool. Identification of soil types was carried out using different empirical diagrams and compared with laboratory classification. The applicability of existing empirical CPT-SPT correlations was examined and a site-specific correlation was proposed. The results show that the available empirical charts work well for the majority of soils. Nonetheless, a refinement is required for the transition zones. Robust correlations between the cone resistance of CPT and the N-value of SPT are still missing and available approximations should be applied with caution.

Characterization of a dense deep offshore sand with CPT and shear wave velocity profiling

J.G. Parra, A. Veracoechea & N. Vieira
Geohidra, Caracas, Venezuela

ABSTRACT: In some areas of the northern South American continental shelf, 89 m below the sea floor a stratum of dense to very dense siliciclastic sand has been identified. Given the properties of the overlying soil layers, this sand is the most competent foundation strata for offshore jacket structures. The proper geomechanical characterization of this sand poses some challenges, among them: the maximum pushing capacity available in the wireline CPT system used in offshore geotechnical exploration, and the extent of the perturbed zone beneath the end of the borehole. This work describes an attempt to overcome those limitations by means of the use of shear wave velocity (Vs) profiling with a P-S logging tool, together with known correlations between Vs and CPT point resistance (qt) to generate more reliable "virtual" CPT records. A MonteCarlo simulation was performed with the objective of assessing the sensitivity of the axial capacity of the proposed foundation piles to the derived CPT records.

Calibration of cone penetrometers in accredited laboratory

J. Peuchen, D. Kaltsas & G. Sinjorgo
Fugro, Nootdorp, The Netherlands

ABSTRACT: Calibration of a penetrometer is an important step in the metrological confirmation of cone penetration testing. This paper summarises experience of Fugro's calibration laboratory based in The Netherlands. This laboratory performs calibrations for about 3000 cone penetrometers used worldwide. The calibration laboratory is accredited to ISO/IEC 17025:2005 "general requirements for the competence of testing and calibration laboratories". This accreditation status is unique for cone penetration test equipment and implies rigorous quality standards.

Defining geotechnical parameters for surface-laid subsea pipe-soil interaction

J. Peuchen
Fugro, Nootdorp, The Netherlands

Z. Westgate
Fugro, Houston, USA

ABSTRACT: The interaction between surface-laid pipelines and the seabed typically takes place within the upper 0.5 m of the seabed. The key geotechnical parameters for assessing pipe-soil interaction are intact and remoulded undrained shear strength, interface shear resistance, submerged soil unit weight, and coefficient of consolidation. This paper summarises experience deriving design geotechnical parameters for various shallow stratigraphy conditions characteristic of different geographic regions. It is shown that characterisation of the upper 0.5 m of the seabed can require site-specific evaluations and ranking of geotechnical investigation tools. This is illustrated for (1) stratigraphic definition and (2) geotechnical parameterisation. The ranking includes cone penetration tests, T-bar/ball penetrometers, free fall penetrometers, box core sampling and in-box miniature T-bar or ball penetration testing, piston core sampling, push sampling and laboratory testing.

Shallow depth characterisation and stress history assessment of an over-consolidated sand in Cuxhaven, Germany

V.S. Quinteros & T. Lunne
Norwegian Geotechnical Institute, NGI, Oslo, Norway

L. Krogh & R. Bøgelund-Pedersen
Ørsted Wind Power (formerly DONG Energy Wind Power), Copenhagen, Denmark

J. Brink Clausen
Geo, Copenhagen, Denmark

ABSTRACT: A comprehensive field campaign was carried out on an aged, dense, over-consolidated sand in Cuxhaven, Germany. CPTs along with a suite of additional in-situ tests, including DMTs, PLTs, shear wave velocity measurements (SDMT and MASW), nuclear densometer, manual drive-in cylinder, temperature, suction and volumetric water content testing were performed on a clean sand to a depth of about four meters. Aim of this field-testing program was to characterize *in-situ* a clean North-Sea sand and by doing so overcome calibration chamber effects (like reconstitution, ageing, and imposed boundary conditions). The field campaign forms part of an effort for the determination of *in-situ* stress history as well as engineering strength and stiffness properties within the 'low' stress regime (< 50 kPa vertical effective stresses). Details of the field results together with an assessment of OCR and K_0 values are introductory presented in this paper. The background of this study is the soil characterisation at 'shallow depths' for subsea structures such as suction buckets, cables, pipelines, satellites and light gravity base structures.

Assessment of pile bearing capacity and load-settlement behavior, based on Cone Loading Test (CLT) results

Ph. Reiffsteck
IFSTTAR, Paris, France

H. van de Graaf
Lankelma, Oirschot, The Netherlands

C. Jacquard
Fondasol, Avignon, France

ABSTRACT: The Cone Loading Test (CLT) consists of stopping the static penetration at a desired level and carrying out a loading of the cone by successive steps. This test is carried out with a standard static penetrometer, equipped with some extra features. The test principle is that at selected depths an electric CPT cone tip is submitted to incremental loading while measuring the settlement of the cone. The test result is a load-displacement curve for both friction sleeve and cone tip. Considering that the cone of the penetrometer is like a reduced model pile, the CLT test is a suitable tool for foundation design. In this paper, a direct method is proposed for using the cone resistance and limit sleeve friction of the CLT test to calculate the bearing capacity and to predict the settlement of a pile at different loads. This paper presents a new approach for transforming a CLT-load-displacement curve into load-displacement curves of a pile (t-z curves). A direct and clear distinction of the bearing capacity of the pile tip and the pile shaft mobilized at different loads becomes possible.

CPT based settlement prediction of shallow footings on granular soils

J. Rindertsma & W.J. Karreman
Van Oord Dredging and Marine Contractors, Rotterdam, The Netherlands

S.P.J. Engels
Arcadis Nederland BV, Arnhem, The Netherlands

K.G. Gavin
Delft University of Technology, Delft, The Netherlands

ABSTRACT: Van Oord DMC executed a large land reclamation project, where requirements had to be met concerning the settlement of shallow foundations to be placed on the sand fill. CPT based settlement predictions performed in the design phase had to be verified with zone load testing. After completing the project, all measurement data were analysed to gain insight into the accuracy of existing correlations between CPT data and settlement of a sand fill. The correlations by De Beer & Martens (1957), Schmertmann (1978), Peck et al. (1996) and Robertson (1990) were considered. From the total number of zone load tests, 43 test were selected which allowed for comparison of the results with the predictions using the mentioned correlations. It was concluded that the CPT to stiffness correlation of Robertson combined with the analytical model of Schmertmann corresponds very well with the measurements, consistently showing only small deviations from the measured settlement.

Analysis of acceleration and excess pore pressure data of laboratory impact penetrometer tests in remolded overconsolidated cohesive soils

R. Roskoden, A. Kopf, T. Mörz & S. Kreiter
Centre of Marine Environmental Science (MARUM), University Bremen, Bremen, Germany

ABSTRACT: Free fall *CPTu* measurements have two data acquisition approaches: direct measurement of sleeve friction (f_s) and tip resistance (q_c) or estimation of these via acceleration data. However, enhanced impact velocities cause strain rate effects on the sediment, which requires to correct q_c, and f_s for these rate effects. The pore pressure (u_2), an important parameter for evaluating soil properties, is not corrected for the penetration rate effects yet. Hence, a laboratory free fall calibration test, with a miniature cone penetration lance, which measures u_2 and acceleration, was designed. The objective was to investigate responses of the acceleration—data to calculate kinematic f_s and q_c. Additionally, we studied dynamic u_2-responses for their penetration rate effect. We also investigated existing penetration rate-corrections for remolded, overconsolidated cohesive soils. In total six experiments were conducted, each of which consisted of four penetrations with impact velocities of 0.02 to 2 m/s. We estimated f_s and q_c from acceleration-data for a remolded, overconsolidated cohesive soil. New rate correction values have been found for this soil, which show a dependency on consolidation histories. Moreover, we identified possible rate effects on u_2-responses as increases in the modulus excess pore pressure. This will help to analyze existing and future u_2-measurements.

CPT-based parameters of pile lengths in Russia

I.B. Ryzhkov
BGAU, Ufa, Russia

O.N. Isaev
Gersevanov NIIOSP, Moscow, Russia

ABSTRACT: In Russia, Cone Penetration Testing (CPT) is developing towards the use of heavy cone penetration rigs to evaluate soil conditions and in particular, calculate the bearing capacity of driven piles. This trend was formed in the former USSR as early as fifty years ago, when mass construction was reoriented toward wide application of precast reinforced concrete and driven piles. The parameters of pile foundations were developed based on the CPT data. In subsequent decades, the range of tasks to be solved by means of CPT significantly expanded, calculations were more reliable and CPT procedures improved.

Comparison of settlements obtained from zone load tests and those calculated from CPT and PMT results

A. Sbitnev
Menard Middle East, Dubai, UAE

B. Quandalle
Menard, Rueil-Malmaison, France

J.D. Redgers
Consultant, UK

ABSTRACT: This paper presents a comparison of settlement results for a construction site where soil improvement by compaction methods was undertaken. A direct comparison of the settlements obtained during the execution of several Zone Load Tests (ZLT) with the settlements estimated based on the results of Cone Penetration (CPT) and Pressuremeter Testing (PMT) for the improved soil is shown. The post treatment PMT and CPT were carried out beyond the depth of the soil treatment. The associated calculations were necessary to predict the settlement of the zone load test adjacent to the test positions. The study addresses an important difference in the results often obtained from ZLT and the calculations performed prior to the execution of the ground improvement work. It also presents the actual site test results and compares them with the calculated settlement based on the comprehensive post treatment testing regime and shows the range of differences in predicted and actual settlements.

Direct use of CPT data for numerical analysis of VHM loading of shallow foundations

J.A. Schneider
USACE, St. Paul, MN, USA

J.P. Doherty
The University of Western Australia, Crawley, WA, Australia

M.F. Randolph
Centre for Offshore Foundation Systems, The University of Western Australia, Crawley, WA, Australia

K. Krabbenhøft
University of Liverpool and Optum CE, Liverpool, England, UK

ABSTRACT: CPT data are often used directly for geotechnical analyses, such as for pile capacity. Advances in automated mesh generating strategies now allow for direct use of CPT data within finite element analyses. This paper presents the results of finite element analyses of shallow foundations with widths varying between 0.5 m and 30 m resting on an offshore soft clay having strength linearly increasing with depth below a crust. Results based on a soil profile using CPT data directly input into the finite element soil model at depth increments of 25 mm are compared with results based on a soil profile with strength linearly increasing with depth below a crust with constant strength. The effects of mesh density and use of remeshing strategies on model run time and results are discussed. Results indicate that numerical analyses based on direct incorporation of a CPT profile may allow for identification of failure surfaces that result in lower optimized capacities than those that try to fit strength using simpler soil profiles.

Evaluation of existing CPTu-based correlations for the undrained shear strength of soft Finnish clays

J. Selänpää, B. Di Buò, M. Haikola & T. Länsivaara
Tampere University of Technology, Tampere, Finland

M. D'Ignazio
Norwegian Geotechnical Institute, Oslo, Norway

ABSTRACT: The Tampere University of Technology has been carrying out an extensive research program on soil testing in Finland. The aim of this research project is to collect data from high-quality in situ and laboratory tests and derive correlations for strength and deformation properties specific to Finnish clays. Correlation models for the undrained shear strength of soft clays based on CPTu measurements have been proposed in the literature by several authors. However, such models are often calibrated from a specific site or soil type. Thus, validation of these models is required before applying them to different soil conditions. In this paper, the existing correlations for the undrained shear strength of soft clays based on CPTu data are compared to test results from different sites in Finland. The validity of the existing models is assessed for Finnish clays by evaluating their bias and uncertainties.

Applications of RCPTU and SCPTU with other geophysical test methods in geotechnical practice

Z. Skutnik, M. Bajda & M. Lech
Warsaw University of Life Sciences, Poland

ABSTRACT: The resistivity cone penetration test (RCPTU) and seismic cone penetration test (SCPTU) results may be used as well for quantitative and for qualitative analysis of the subsoil on tested sites. Electrical resistivity tomography (ERT) with RCPTU and SCPTU procedures are capable of detecting discrete horizons that would normally be missed using other tests at specific depth intervals. In the first tested site the RCPTU and SCPTU results have been used to define sub-surface stratigraphy and the soil porosity of the Pliocene clay deposits, while in the other site to identify the layers of organic subsoil and deformation parameters of subsoil. In the third tested site geotechnical in situ investigations were carried out using of 10 cm^2 CPTU cone and 15 cm^2 RCPTU cone. All these tests may be carried out in the preliminary investigation phase of ground conditions as well as supplementary studies and are most effective when are combined.

CPT in a tropical collapsible soil

C.S.M. Soares, F.A.B. Danziger, G.M.F. Jannuzzi & I.S.M. Martins
Federal University of Rio de Janeiro, Brazil

M.E.S. Andrade
Technological Federal University of Paraná, Brazil

ABSTRACT: A test site was established in Primavera do Leste, a city 200 km east of Cuiabá, Central-West region of Brazil, one of the main Brazilian areas of soya and corn production. The research aimed at improvement of the knowledge of soil characteristics in that region, where serious problems related to foundations of silos are frequent. The present paper presents cone penetration tests performed in dry and rainy seasons. One block sample and tube samples have been collected, and characterization, determination of clay minerals and oedometer tests in specimens in natural water content and on soaked specimens have been performed. Soil collapsibility of the upper layer was verified. CPT results indicated three soil behaviour types, however for one of those layers the soil behaviour type was not properly indicated by the CPT. The CPTs performed in dry and rainy seasons allowed the conclusion that the variation of the water content is limited to 2.0–2.5 m depth.

Liquefaction resistance by static and vibratory cone penetration tests

F.T. Stähler, S. Kreiter, M. Goodarzi, D. Al-Sammarraie & T. Mörz
MARUM-Center for Marine Environmental Sciences, University of Bremen, Bremen, Germany

ABSTRACT: Soil liquefaction is an important hazard and one of the causes of earthquake-related disasters. The prediction of soil liquefaction is a major issue in geotechnical engineering and the soil liquefaction resistance is often determined in-situ by static Cone Penetration Tests (CPT). The determination of the soil liquefaction resistance relies on empirical correlations and one way to increase accuracy might be the use of vibratory CPT. We report on displacement-controlled vibratory and static CPT performed in a calibration chamber under a simulated field boundary condition, known as BC5. Vibratory CPT led to a reduction in cone resistance for medium- to very-dense Ticino sand. The reduction ratio increased at high vibrational frequencies and was independent of the relative density for the specific stress state and type of soil. A correlation between the liquefaction resistance and static or vibratory CPT is proposed.

In situ characterisation of gas hydrate-bearing clayey sediments in the Gulf of Guinea

F. Taleb, S. Garziglia & N. Sultan
IFREMER, Département REM, Unité des Géosciences Marines, Plouzané, France

ABSTRACT: Increasing needs for energy resources have moved deep offshore developments and research efforts towards regions where high pressure and low temperature conditions allow gas and water to form Gas Hydrates (GH). However, owing to difficulties in sampling and preserving GH under in-situ stability conditions, GH-bearing sediments remain particularly challenging to characterise using conventional laboratory methods. This paper presents the experience gained in the use acoustic and piezocone soundings in characterising gas-hydrate bearing clayey sediments offshore Nigeria. Acoustic measurements of compressional wave velocity are shown to be an expedient means of both detecting and quantifying GH in sediments. Piezocone data and their location in normalised soil classification charts highlight trends in response suggesting that GH-bearing clayey sediments are predominately contractive at large strains. The observed trend of increasing cone resistance combined with increasing pore pressure suggests that the sensitivity of GH-bearing sediments tend to increase with their stiffness and strength.

Gas effect on CPTu and dissipation test carried out on natural soft-soil of Barcelona Port

D. Tarragó & A. Gens
Universitat Politècnica de Catalunya, Barcelona, Spain

ABSTRACT: There is abundant evidence that gas is present in the sediments of the prodelta of the Llobregat River in the area of the Barcelona Port. The presence of gas has been directly observed by gas releases and deflagrations occurring during a site investigation campaign involving boreholes and CPTu tests. Methane has been identified by means of BAT permeameter tests. The record of a CPTu test performed in an area where gas is known to be present has been examined in detail. In spite that there are several features that could be related to the gas presence, it is challenging to identify unambiguously the presence of the gas from the CPTu record alone.

Comparison of cavity expansion and material point method for simulation of cone penetration in sand

Faraz S. Tehrani & Vahid Galavi
Department of Geo-Engineering, Deltares, Delft, The Netherlands

ABSTRACT: Over the years, many attempts have been made to simulate the cone penetration process in coarse-grained and fine-grained soils. The simulation methods range from pure analytical methods to complicated numerical approaches. Among the analytical methods, cavity expansion analysis has gained more attractions due to its relative simplicity and robust theoretical background. Analytical cavity expansion analysis is limited to rather simplistic soil constitutive models. As a result, numerical cavity expansion analyses have been attempted to cover this limitation. Another method that has gained much attention in the past decade is the Material Point Method (MPM) whose promising capabilities in simulating large deformation problems can be used for simulating the cone penetration. In this paper, we model the cone penetration in loose, medium dense and dense sands using numerical spherical cavity expansion method and axisymmetric MPM. In all the analyses, soil behavior is modelled using Mohr-Coulomb constitutive model. The soil is assumed to be dry in all analyses. Results are compared with a cylindrical cavity expansion solution.

Soil behavior and pile design: Lesson learned from some prediction events—part 1: Aged and residual soils

G. Togliani
Geologist, Massagno, Switzerland

ABSTRACT: Pile design methods generally do not consider phenomena such as aging, cementation and weathering, responsible for the presence of microstructures in the soil surrounding the piles and affecting their performance. Some prediction events involving known structured soils are analyzed taking as a reference the CPT-based SBT classification system recently updated by P.K. Robertson (2016). It is shown that the normalized small-strain rigidity index K^*_G can be useful to assess the impact of soil microstructure on pile capacity.

Soil behavior and pile design: Lesson learned from recent prediction events—part 2: Unusual NC soils

G. Togliani
Geologist, Massagno, Switzerland

ABSTRACT: Pile design methods generally do not consider that the presence in the soil surrounding the piles of a microstructure related to phenomena such as aging, cementation and weathering, could significantly affect their performance. Analyzing soils classified as NC with reference to the CPT-based SBT classification system recently updated by P.K. Robertson, it is possible to highlight, via the normalized small-strain rigidity index K^*_G, the presence of a microstructure and to define his impact on piles design.

A probabilistic approach to CPTU interpretation for regional-scale geotechnical modelling

L. Tonni & M.F. García Martínez
Department of DICAM, University of Bologna, Italy

I. Rocchi
Department of Civil Engineering, Technical University of Denmark, Denmark

S. Zheng & Z.J. Cao
State Key Laboratory of Water Resources and Hydropower Engineering Science, Wuhan University, China

L. Martelli & L. Calabrese
Servizio Geologico, Sismico e dei Suoli, Regione Emilia-Romagna, Italy

ABSTRACT: The paper describes part of a study carried out to develop the geotechnical model of a coastal area on the Adriatic Sea, between the municipalities of Cesenatico and Bellaria-Igea Marina in the Emilia-Romagna region (Italy). A large experimental database, provided by the Geological, Seismic and Soil Survey of the Emilia-Romagna Authority, was used to develop a stratigraphic model of the upper 30 m subsoil of this coastal area, together with estimates of the mechanical parameters of the different soil units. A Bayesian approach was used to identify the most probable number of soil layers and their thicknesses, based on the Soil Behaviour Type Index obtained from CPTU results. This tool has already been used for small scale areas and its implementation in large datasets could eventually provide a preliminary estimate of the expected soil conditions at a site, taking into account statistically the inherent spatial variability in a rational and transparent way.

CPTu-based soil behaviour type of low plasticity silts

L.A. Torres-Cruz
University of the Witwatersrand, Johannesburg, South Africa

N. Vermeulen
Jones & Wagener, Johannesburg, South Africa

ABSTRACT: The application to low plasticity silts of two CPTu-based soil behaviour type charts proposed by Schneider and co-workers during the last decade is explored. One chart is based on tip resistance and sleeve friction (Q-F), and the other is based on tip resistance and dynamic pore water pressure (Q-$U2$). CPTu soundings made at two mine tailings deposits with a predominance of silt-sized particles were considered. The results indicate that the mechanical response of the CPT in these low plasticity silts can be indistinguishable from that of sands. As such, it is argued that the potential for misleading classifications can be reduced if the name of the 'sands' zone of the charts includes low-plasticity soils in general, regardless of fines content. Additionally, a new boundary for the drained zone of the Q-$U2$ chart is proposed to enhance the classification of loose or very compressible low-plasticity soils in which largely drained penetration is possible at low values of Q.

Interpretation of soil stratigraphy and geotechnical parameters from CPTu at Bhola, Bangladesh

Zinan A. Urmi & M.A. Ansary
Bangladesh University of Engineering and Technology, Dhaka, Bangladesh

ABSTRACT: Soils are very complex materials and characterization of soil behavior requires interpretation of many geotechnical parameters, such as soil stratigraphy, stress history and index properties. Piezocone Penetration Test (CPTu) is a proven tool for effective and reliable subsoil exploration and for evaluation of near continuous soil stratigraphy and interpretation of many geotechnical parameters. The CPTu measurements (q_c, f_s and u_2) can be interpreted by using theoretical, analytical, and statistical methods; or by using other in-situ tests techniques or laboratory investigations to evaluate suitable geotechnical parameters that can be used in engineering practice. This paper presents interpretation of soil stratigraphy and evaluation of geotechnical parameters from CPTu measurements using laboratory investigations and in-situ Standard Penetration Test (SPT). Three pairs of CPTu and SPT tests were performed on the coastal embankment located in southern part of Bangladesh. The assessed geotechnical parameters include Soil Behavior Type, Mean Grain Size and Fines content. It is expected that this paper will provide valuable insight of CPTu data interpretation of the studied geologic setting, at Bhola, Bangladesh.

Thermal Cone Penetration Test (T-CPT)

P.J. Vardon
Geo-Engineering Section, Delft University of Technology, The Netherlands

D. Baltoukas & J. Peuchen
Fugro, The Netherlands

ABSTRACT: The Thermal Cone Penetration Test (T-CPT) records temperature dissipation during an interruption of the Cone Penetration Test (CPT) to determine the thermal properties of the ground, taking advantage of heat generated in the cone penetrometer during normal operation. This paper compares two interpretation models for thermal conductivity. It is found that the thermal conductivity can be accurately determined. Care must be taken of the initial heat distribution and sensor location within the temperature cone to achieve accurate results. Furthermore, laboratory test data are presented that show that the full-displacement push of a penetrometer into sandy strata has limited influence on thermal conductivity values.

Development of numerical method for pile design to EC7 using CPT results

J.O. Vasconcelos, J. O'Donovan & P. Doherty
Gavin and Doherty Geosolutions Ltd., Dublin, Ireland

S. Donohue
University College Dublin, Dublin, Ireland

ABSTRACT: In the UK and Ireland bearing pile design is generally based on the "design by calculation" methodology where soil strength parameters are obtained from in-situ data or through empirical correlation with ground test results. Direct ground test results from Cone Penetration Tests (CPT's) are seldom used to complete ultimate limit state designs to satisfy EC7. This paper discusses the development of a numerical method where direct results from CPTs are used to design bearing piles. The method was verified through a case study where a large amount of CPT data was available. The resulting pile designs were compared to the more commonly used "design by calculation" approach and "design by pile load test". The numerical method was incorporated into the AllPile (Axially and Laterally Loaded Pile) software which supports single pile design according to EC7.

Prehistoric landscape mapping along the Scheldt by camera- and conductivity CPT-E

J. Verhegge & Ph. Crombé
Department of Aarchaeology-Prehistory Research Unit, Ghent University, Gent, Belgium

M. van den Wijngaert
Geosonda Environment nv, Sint-Denijs-Westrem, Belgium

ABSTRACT: Over the past decade, paleolandscape reconstruction was introduced as part of a preventive archaeological evaluation strategy along the Scheldt river, due to the unexpected discovery of well-preserved prehistoric landscapes and sites during construction works in the Antwerp harbor area. Hereby, CPT is an important tool in combination with coring and/or near surface geophysical survey. Applications of CPT range from desktop studies, which determine evaluation strategies, to actual paleolandscape mapping by sedimentological data interpretation. CPT-Es (with added camera and/or electrical conductivity sensors) calibrate and validate geophysical subsurface modelling and soil behavior types are interpreted. Particularly, (electrical conductivity) CPT-C disentangles sedimentological and hydrological variations in electrical conductivity values. On the other hand, camera CPT improves differentiation of organic rich sediments and detection of thin organic soil horizons within homogenous (cover)sands. The usability of CPT is illustrated through recent prehistoric landscape evaluation studies along the Scheldt river.

Comparative analysis of liquefaction susceptibility assessment by CPTu and SPT tests

A. Viana da Fonseca, C. Ferreira, A.S. Saldanha & C. Ramos
CONSTRUCT-GEO, Faculty of Engineering, University of Porto, Porto, Portugal

C. Rodrigues
Polytechnic Institute of Guarda, Guarda, Portugal

ABSTRACT: The assessment of liquefaction susceptibility from field tests is conventionally based on the Factor of Safety (FS_{liq}) against liquefaction, relating the Cyclic Resistance Ratio (CRR) with the Cyclic Stress Ratio (CSR). The calculation of CSR is relatively straightforward, whereas CRR strongly depends on the in situ technique from which it is derived. Distinct approaches have been proposed based on quantitative liquefaction risk indexes, namely the Liquefaction Potential Index (LPI) and the Liquefaction Severity Number (LSN). In Portugal, a pilot site for liquefaction assessment has been set up in the Lower Tagus Valley, near Lisbon, within the European H2020 LIQUEFACT project. In this paper, the geotechnical field data from SPT and CPTu is integrated in the three approaches to liquefaction assessment. A comparative analysis of the results is presented and discussed, highlighting the differences and limitations of these in situ tests in the assessment of liquefaction susceptibility in loose granular soils.

Application of CPT testing in permafrost

N.G. Volkov & I.S. Sokolov
GEOINGSERVICE (Fugro Group), Moscow, Russia

R.A. Jewell
Fugro GeoConsulting, Brussels, Belgium

ABSTRACT: This paper introduces some of the physical processes influencing the mechanical properties of permafrost and the effect of temperature change in the range 0°C to −10°C. The critical role of saline concentration in the pore fluid is described. Examples of recent CPT investigations at different permafrost sites in Russia are given to highlight the type of data that can be obtained and the interpretation for foundation engineering. Some of the investigations were completed in extremely cold conditions.

Comparison of mini CPT cone (2 cm^2) vs. normal CPT cone (10 cm^2 or 15 cm^2) data, 2 case studies

G.T. de Vries, C. Laban & E. Bliekendaal
Marine Sampling Holland, Velsen Noord, The Netherlands

ABSTRACT: For decades, the standard for CPT testing has pre-scribed use of a cone with a surface area of 10 cm^2. For cones with a projected surface area smaller than 5 cm^2, standards appear to question and suggest the application of a correction factor.

This paper discusses results of CPT tests performed with a Mini CPT system, in which cones with a nominal surface area of 2 cm^2 were used, compared to CPT results obtained with a standard 10 cm^2 or 15 cm^2 CPT cone. Two cases are presented. In case 1, project objective was validation of mini cone results for further use of the mini CPT system during construction works at the Dutch Maasvlakte 2 port extension. In the other case, the mini CPT system was used to proof soil deterioration along a newly built quay in a Dutch port. Finally, some suggestions are made regarding improvements to- and application of Mini CPT systems.

The development of "Push-heat", a combined CPT-testing/thermal conductivity measurement system

G.T. de Vries
Marine Sampling Holland, Velsen Noord, The Netherlands

R. Usbeck
Fielax, Bremerhaven, Germany

ABSTRACT: In design of offshore powercables, thermal characteristics of the soil in which the future cable will be located are needed to calculate and predict cable capacity and lifetime. Traditionally, thermal characteristics were measured in the laboratory, on samples obtained with various drilling and coring techniques. Unfortunately, for various reasons, results of laboratory testing are not always consistent. This necessitated the design of a reliable system to measure thermal characteristics, both in the laboratory and in-situ. Older in-situ measurement techniques have mainly focussed on deployment in soft soil and deep water. For use in coastal environments, with complicated soil profiles and soil types more resilient to penetration, these older systems have only limited practical use. Early development of a system combining vibrocoring with in situ thermal conductivity measurement has proved reasonably succesfull but in some soil types still rendered anomalous results. This paper discusses the development of a new measurement system which combines traditional CPT pushing- and measurement technology with an existing well established in-situ thermal measurement technique. Some initial measurement results are presented.

Keywords: CPT testing, thermal conductivity, in-situ testing, equipment, testing and procedures

Free fall penetrometer tests in sand: Determining the equivalent static resistance

D.J. White
University of Southampton, UK

C.D. O'Loughlin
University of Western Australia, Australia

N. Stark
Virginia Tech, USA

S.H. Chow
University of Western Australia, Australia

ABSTRACT: Free Fall Penetrometer (FFP) tests provide an efficient way to determine the penetration resistance at shallow depths in sandy soils, and are being used increasingly in geotechnical, geomorphological and coastal engineering applications. A limitation of free fall penetrometers is the effect of their high velocity on the penetration resistance. This affects the drainage condition, creates a viscous-type enhancement of the mobilised strength, and also introduces inertial drag forces. It is useful if the measured FFP resistance can be adjusted back to the resistance that would be expected in a standard Cone Penetrometer Test (CPT) at the same location. With this adjustment, the resistance can be used in the same correlations and design methods as standard CPT data. Adjustments for viscous-type rate effects and inertial drag have been proposed and explored in detail for clay soils. The contribution of this paper is to outline a correction scheme for drainage condition, which is more relevant for sandy soils. This correction utilizes the dissipation response at the end of the FFP test, in combination with the measured or derived FFP tip resistance. Relationships for penetration resistance in drained and undrained conditions based on density state are developed. It is shown that the high velocity FFP resistance can be uniquely mapped to a resistance from a standard CPT, when combined with the dissipation response. With development and validation, this new framework could enhance the value of FFPs as a complementary or alternative technology alongside conventional static penetration testing.

The variability of CPTU results on the AMU-Morasko soft clay test site

J. Wierzbicki & R. Radaszewski
Institute of Geology, Adam Mickiewicz University, Poznan, Poland

M. Waliński
Geoprojekt-Poznan S.C., Poland

ABSTRACT: The paper presents results of tests conducted on a test site at Adam Mickiewicz University (AMU) in Poznań, Poland. The test site is located in the area of distribution of normally consolidated postglacial clayey silty sands and sandy clays. These soils are characterized by high porosity and are fully saturated, therefore, from a geotechnical point of view they can be defined as soft clays. There were 9 CPTU tests conducted along with research drillings and taking samples for laboratory analyses in the area of 8 ha. This paper focuses on the analysis of the uniformity of the soft clay layer found on the test site and on the analysis of variability of the CPTU tests conducted there. The results obtained enabled to separate a statistically homogeneous batch of soil in the subsoil and to form a reference set of CPTU results characteristic of the soft clay layer.

Shear strengths determined for soil stability analysis using the digital Icone Vane

M. Woollard & O. Storteboom
A.P. van den Berg, Heerenveen, The Netherlands

A.S. Damasco Penna
Damasco Penna, São Paulo, Brazil

É.S.V. Makyama
Damasco Penna, São Paulo, Brazil

ABSTRACT: At locations where the soil is exposed to high and varying forces, for example at a dike or around a mining area, additional parameters for stability analysis should often be measured. Determination of the undrained shear strength is a commonly used method to define soil stability. The Field Vane Test (FVT) can be used for in-situ measurement and evaluation of the shear strength. It can be deployed in soft soils, but also in fine-grained soils such as silts, organic peat, tailings and other geomaterials where a prediction of the undrained and remolded shear strength is required. This paper describes a FVT system called Icone Vane, which is part of the Icone® data acquisition concept that is based on fully digital data transfer and meets the requirements of ASTM D2573/ D2573M and other FVT standards. Specific details of fieldwork with the Icone Vane performed at tailings dams are presented.

Metal objects detected and standard parameters measured in a single CPT using the Icone with Magneto click-on module

M. Woollard & O. Storteboom
A.P. van den Berg, Heerenveen, The Netherlands

L. Gosnell & P. Baptie
1st Line Defence Ltd, Hoddesdon Herts, UK

ABSTRACT: The paper describes a digital CPT system called Icone®. This system is easily extendable by click-on modules to measure additional parameters and any module is automatically recognized by a digital data logger, thus creating a true plug & play system. By moving to smart digital communication, sufficient bandwidth over a thin flexible measuring cable was created to accommodate additional parameters, without the need for changing cones, cables or data loggers. The following click-on modules are described: seismic, conductivity, magneto and vane. Feedback from fieldwork with the Icone and the magnetometer module (Magneto) highlights the user experience with this approach.

Using the magnetometer module, metal objects in the underground can be detected by interpreting anomalies of the earth's magnetic field. The application of this module is illustrated by two unexploded ordnance (UXO) survey projects for clearance ahead of piling.

Simulation of liquefaction and consequences of interbedded soil deposits using CPT data

Fred Yi
Terracon Consultants, Inc., Colton, CA, USA

ABSTRACT: Several corrections have been proposed since 1990s for the interpretation of liquefaction potential and analysis of the consequences of liquefaction based on cone penetration test (CPT) measurements. However, few publications address how these factors affect the calculation results. In this study, extensive studies of soil liquefaction and its consequences were performed using selected CPT data from two sites located within 7 km of the fault rupture of the 1999 Kocaeli, Turkey earthquake ($M_w = 7.5$). The effects of thickness of interval for averaging, thin layer correction and transition zone were examined in detail.

Correlations among SCPTU parameters of Jiangsu normally consolidated silty clays

H. Zou, S. Liu & G. Cai
Southeast University, Nanjing, China

A.J. Puppala
University of Texas at Arlington, Texas, USA

ABSTRACT: The enhanced seismic piezocone penetration testing (SCPTU) provides a convenient method to measure V_s together with other three indices including the cone tip resistance, sleeve frictional resistance and pore water pressure in a close proximity. In this study, a database containing 117 sets of data points is compiled for the stress-normalized SCPTU parameters of the Jiangsu normally consolidated silty clays. An approach based on the multivariate distribution method is utilized to develop the multivariate correlations among the normalized shear wave velocity (V_{s1}) and other three indices. It is shown that the correlations are reliable to describe the dependence of the V_{s1} on other three indices for the Jiangsu normally consolidated silty clays. Moreover, the uncertainties within V_{s1} can be reduced by incorporating more than one index in the prediction.

Author index

Agaiby, S.S. 65
Al-Baghdadi, T. 92
Albatal, A. 77
Alderlieste, E.A. 117
Allievi, L. 66
Al-Sammarraie, D. 67, 144
Amoroso, S. 68, 90
Anamali, E. 69
Andrade, M.E.S. 91, 143
Ansary, M.A. 152
Arroyo, M. 87, 103, 124
Augarde, C. 92

Bagińska, I. 71
Bajda, M. 142
Bałachowski, L. 72, 110
Ball, J. 92
Baltoukas, D. 153
Baptie, P. 163
Barbosa, H.T. 88
Barounis, N. 73, 74
Bates, L. 107
Baziw, E. 75
Bienen, B. 3, 86
Bihs, A. 76
Bilici, C. 77
Blake, A. 92
Bliekendaal, E. 158
Bøgelund-Pedersen, R. 134
Boone, M.D. 78
Boulanger, R.W. 25
Bray, J.D. 79
Brennan, A. 92
Brik, A. 80, 81
Brink Clausen, J. 134
Brizzolara, S. 125
Brown, M. 92

Cai, G. 165
Calabrese, L. 150
Calvello, M. 98
Camacho, C.B. 83
Camacho, M.A. 83
Cao, Z.J. 150
Cardoso, A. 84
Carlson, M. 118
Carroll, R. 85

Chow, S.H. 3, 86, 160
Christopher, N. 99
Chung, S.G. 126
Collico, S. 87
Coombs, W. 92
Coutinho, R.Q. 88, 89
Crombé, Ph. 155
Cruz, J. 90
Cruz, M. 90
Cruz, N. 90
Cubrinovski, M. 79
Cuomo, S. 98
Curran, T. 66

Damasco Penna, A.S. 162
Danziger, F.A.B. 91, 143
Davidson, C. 92
Davidson, J. 84
de Gast, T. 97
de Greef, J. 101, 113
de la Torre, C. 79
de Lange, D.A. 112
De Nijs, G.A. 70
De Nijs, G.J.J. 70
de Vries, G.T. 158, 159
Deakin, R. 66
DeJong, J.T. 25
Devincenzi, M. 87
Dhimitri, L. 69
Di Buò, B. 82, 141
D'Ignazio, M. 82, 114, 141
Dmitriev, G.Y. 108
Doan, L.V. 93, 94, 95
Doherty, J.P. 3, 140
Doherty, P. 154
Donohue, S. 154
Duffy, W.P. 78

Engels, S.P.J. 136

Ferreira, C. 156
Fityus, S. 107
Flores-Eslava, R. 106

Galavi, V. 98, 147
García Martínez, M.F. 96, 150

Garziglia, S. 145
Gavin, K.G. 45, 111, 136
Gens, A. 124, 146
Gerken, D.E. 78
Ghasemi, P. 98
Gibbs, P. 99
Goodarzi, M. 67, 100, 144
Gosnell, L. 163
Gottardi, G. 96
Gundersen, A.S. 102, 114
Gusmão, A.D. 88
Gylland, A. 76

Haikola, M. 141
Haugen, E. 118
Hauser, L. 103
Hayashi, H. 104
Helle, T.E. 105
Heredia, W. 83
Hermann, S. 114
Hicks, M.A. 97

Ibarra-Razo, E. 106
Imre, E. 107
Isaev, O.N. 108, 109, 138

Jacquard, C. 135
Janecki, W. 71
Jannuzzi, G.M.F. 91, 143
Jewell, R.A. 157
Joosten, S. 113

Kaltsas, D. 132
Karabacak, D. 116
Karlsrud, K. 114
Karreman, W.J. 136
Kåsin, K. 118
Kassner, M. 118
Kawa, M. 71
Kleven, A. 114
Kluger, M.O. 100
Knappett, J. 92
Konkol, J. 72, 110
Kopf, A. 115, 137
Kovacevic, M.S. 111
Krabbenhøft, K. 140
Krall, T. 79

Kreiter, S. 67, 100, 137, 144
Krogh, L. 99, 134

Laban, C. 158
Länsivaara, T. 82, 141
Lech, M. 142
Lehane, B.M. 93, 94, 95
Lengkeek, H.J. 101, 113
Lewis, M.R. 78
L'Heureux, J.S. 102, 114, 115, 118
Libric, L. 111
Liu, S. 165
Lo Presti, D.C. 121
Long, M. 76, 105, 115
Looijen, P. 116
Lundberg, A.B. 117
Lunne, T. 91, 102, 118, 134

Makyama, É.S.V. 162
Mantaras, F.M.B. 127
Marchetti, D. 68
Martelli, L. 150
Martinelli, M. 98
Martins, I.S.M. 143
Mayne, P.W. 65, 119, 129
McCallum, A.B. 120
McNinch, J.E. 77
Meisina, C. 121
Meissl, S. 84
Mejia, J. 83
Mellia, F.C. 89
Międlarz, K. 72
Minarelli, L. 68
Minkin, M.A. 108
Młynarek, Z. 122
Mo, P.Q. 123
Mohr, H. 3
Monaco, P. 68
Monforte, L. 103, 124
Mörz, T. 67, 100, 137, 144
Mumtaz, M.B. 125

Nguyen, T.D. 126
Nielsen, S.W. 99
Nordal, S. 76, 105

Odebrecht, E. 127
O'Donovan, J. 154
O'Loughlin, C.D. 3, 160
Orazalin, Z.Y. 128
Ouyang, Z. 129

Paniagua, P. 76, 85, 114
Papamichael, S. 130
Parasie, N. 116

Parolini, C. 124
Parra, J.G. 131
Pedersen, R.B. 99
Perez, N. 87
Peuchen, J. 116, 119, 132, 133, 153
Philpot, J. 73, 74
Pinheiro, A.V.S. 91
Powell, J.J.M. 69
Puppala, A.J. 165

Quandalle, B. 139
Quinteros, S. 102
Quinteros, V.S. 134

Radaszewski, R. 161
Ragni, R. 3
Ramos, C. 156
Randolph, M.F. 3, 86, 140
Rangel-Núñez, J.L. 106
Raymackers, S. 84
Reale, C. 111
Redgers, J.D. 139
Reiffsteck, Ph. 135
Rica, S. 70
Richards, D. 92
Riemens, H.J. 70
Rindertsma, J. 136
Rivera-Cruz, I. 106
Robertson, P.K. 78, 80
Rocchi, I. 96, 150
Rodrigues, C. 90, 156
Rømoen, M. 114
Roskoden, R. 137
Rouainia, M. 100
Ryzhkov, I.B. 108, 138

Saldanha, A.S. 156
Salinas, L.M. 83
Sampurno, B. 99
Sbitnev, A. 139
Schanz, T. 107
Schnaid, F. 127
Schneider, J.A. 3, 140
Schneider, M.A. 3
Schweiger, H.F. 103
Selänpää, J. 82, 141
Sharafutdinov, R.F. 108, 109
Sinjorgo, G. 132
Skutnik, Z. 142
Smaavik, T. 114
Soares, C.S.M. 143
Sokolov, I.S. 157
Stacul, S. 121
Stähler, F.T. 67, 100, 144
Stanier, S.A. 3

Stark, N. 77, 125, 160
Stefani, M. 68
Stefaniak, K. 122
Stocks, E. 79
Storteboom, O. 162, 163
Strandvik, S. 118
Sultan, N. 145

Taleb, F. 145
Tarragó, D. 146
Tehrani, F.S. 147
Terwindt, J. 112
Togliani, G. 148, 149
Tonks, D. 81
Tonni, L. 96, 150
Torres-Cruz, L.A. 151
Tyldsley, C. 66

Urmi, Z.A. 152
Uruci, E. 118
Usbeck, R. 159

Van Baars, S. 70
van de Graaf, H. 135
van den Wijngaert, M. 155
van der Linden, T.I. 112
Vanneste, M. 115
Vardon, P.J. 97, 153
Vasconcelos, J.O. 154
Veldhuijzen, A. 118
Veracoechea, A. 131
Verbeek, G. 75
Verhegge, J. 155
Vermeulen, N. 151
Viana da Fonseca, A. 156
Vieira, N. 131
Volkov, N.G. 108, 157
Vrettos, C. 130

Wadman, H. 77
Waliński, M. 161
Wang, L. 92
Ward, D. 69
Westgate, Z. 133
White, D.J. 3, 160
Whittle, A.J. 128
Wierzbicki, J. 122, 161
Woollard, M. 162, 163

Yamanashi, T. 104
Yi, F. 164
Yu, H.S. 123

Zakatov, D.S. 109
Zheng, S. 150
Zou, H. 165

PGMO 06/07/2018